Antenna Designs for NFC Devices

Antenna Designs for NFC Devices

Dominique Paret

ISTE

WILEY

First published 2016 in Great Britain and the United States by ISTE Ltd and John Wiley & Sons, Inc.

ISTE Ltd
27-37 St George's Road
London SW19 4EU
UK

www.iste.co.uk

John Wiley & Sons, Inc.
111 River Street
Hoboken, NJ 07030
USA

www.wiley.com

Library of Congress Control Number: 2015955805

British Library Cataloguing-in-Publication Data
A CIP record for this book is available from the British Library
ISBN 978-1-84821-841-3

Contents

Acknowledgments

During my long career in three disciplines – first as a professional and industrialist at Philips/NXP Semi-Conductors, then and simultaneously as a teacher at numerous engineering schools and universities, and finally, for the past 10 years, as a founder of dp-Consulting (a firm of consultants and independent technical experts) – I have had the opportunity to meet many experts in this field. Therefore, it is extremely difficult to individually thank everyone to whom thanks are due – that could fill an entire book and more. In addition, as near-field communication (NFC) has such a vast range of applications, it would be fallacious to try to write such a book as this on the subject alone, so my heartfelt thanks go to numerous colleagues and friends:

– from NXP Semi-Conductors in Graz (Austria), Hamburg (Germany), Monza (Italy) and Caen (France), with whom I have worked for many years;

– customers, partners and competitors whom I frequently meet at working meeting of the ISO and “mirror” commissions at AFNOR (France’s standardization body).

In addition, in the coming chapters, I will occasionally address some specific acknowledgments to various friends for their help.

I dedicate this book to the many, many readers who have followed me over the years, through series of publications and technical reports. I owe to you many of these words of gratitude, because it is due to you and for you that I have been able to take my courage in both hands to squeeze as much detail out of the subject as possible, with a view to presenting some of its mysteries in as didactic a manner as possible. In any case, thanks again to one and all for your help and your assiduousness in reading my work. If these few words of thanks have encouraged you to persevere in learning techniques and given you a vocation as a trainer and pedagogue, I would be delighted.

Once again, my true thanks to all of you for your contributions and your faithful friendship.

Dominique PARET
November 2015

Preface

Why this book?

Having worked in the field of near-field communication (NFC) since its very beginning (a little over 15 years now), I have a great deal of written material about this subject. Many advances have been made on the basis of the operational principles, and wonderful and highly varied applications (*software*) in NFC (particularly with mobile phones). However (there is always a *but*), there has been relatively little detailed literature written on the specific functions and the mysteries of application-specific integrated circuits and antennas (*hardware*) involved in NFC, and even less about the physical layers, i.e. the air interface. It is my hope that this book will go a long way toward filling this void.

Therefore, I would rather clarify the situation immediately. It is not because of pure ideology that I wish to rectify this shortcoming, but instead because it is a day-to-day reality. Once software bugs have been worked out and, on paper, the application should work, the sticking point is the physical part – the lower layers in the open systems interconnection (OSI) model – which designers and users have blithely overlooked, saying “Everything will be alright!”. Unfortunately, communication in the air interface, with the antennas that involve the numerous form factors of the different applications and the physical environments, presents us with problems and grim realities which need to be seriously considered, without which nothing can be done. I have often said throughout my career that “there is no point singing the praises of any piece of software, no matter how good it is, if the motherboard doesn’t work!”. This is another way of saying “Don’t put the cart before the horse!”.

For whom?

Having been a technical consultant for many years to numerous companies, navigating in mysterious waters, and in view of the thriving NFC market, I have designed this book to serve engineers, technicians, students and the growing number of new arrivals to this domain, to help them avoid certain pitfalls. At present, there is little fundamental information (both basic technical and applicational training) available on these subjects. Hence, today, I offer this book to give you an initiation in the field and a specific knowledge of the techniques of design of antennas devoted to NFC devices, and I have taken it upon myself to gather, compile and collate a technical database, made up of various generic industrial examples, as well as concrete and highly detailed examples of "antennas for NFC Devices… and possibly more", with widespread involvement from my company, dp-Consulting.

The aim of this book is not to be a coverall in marketing, and a joyous ramble through the subject, but instead a technical reference offering an in-depth examination of all the technical details and all the functional and application issues pertaining to these technologies. As readers will discover over the course of the book, the field of applications and usages is enormously varied – ranging from the exchange of dematerialized data to secure payment, controlled by a pocket device (such as a mobile phone and a camera), which can also be used as a chip card or a reader, or a television in a car, etc. Thus, we have a multitude of "form factors", environments, etc., leading to a multitude of technical problems where antennas are concerned.

Additionally, as per usual, so as not to hamper readers' understanding of the devices presented herein, I have done my utmost to make the book as clear and instructive as possible, so they can instantly see the point being made, and the link between theory, technological aspects, financial factors, etc.

For now, I hope that readers will find this book enjoyable and illuminating. Above all, enjoy it, because it is for you (not for me) that I have written it. On that point, if any shadow of a doubt remains, readers are always welcome to contact me, putting forward any comments, remarks, questions, etc. (provided that they are constructive, of course) about the content and form of this book, via e-mail, at dp-consulting@orange.fr.

IMPORTANT NOTES.– It must also be noted that this book is intended to complement many of my other works. Those works offer a more specific, even more detailed focus on the techniques used in contactless technology, in near fields in high frequency (HF) and NFC, and the implementation of those techniques. This further reading should also state the technical curiosity of the vast majority of users. Note, also, that I have borrowed and reproduced from those books[1] a number of sections written previously, because it is very difficult not to simply rewrite the same things on the same subjects, all the more so given that those writings on the physical fundaments of NFC date from the late 19th Century (with Biot, Savart, Maxwell and Laplace). Given that this field is evolving rapidly, it will inevitably be necessary to update the content of these books in 3 or 4 years' time, but in any case, until then, at least readers will have the fundamental principles in place.

Warning

This book is not intended to be a literal reproduction of the scant application notes that anyone can (with a certain amount of difficulty) find on the Web, but is instead meant to serve as an introduction to, and a detailed, instructive presentation of the principles and technical workings of NFC, and also to give people coming into this field an overall conceptual and applicational view of it.

EXTREMELY IMPORTANT NOTE.– From the very start, I wish to draw readers' attention to the important fact that, in order to give adequate coverage of the field of NFC, this book describes a great many patented technical principles, subject to the holding of licenses and their associated rights (bitcodes, communication techniques, etc.). They have already been published in official professional technical texts/communications, or at conferences/public seminars. But, above all, they must be used in accordance with the legislation in force.

1 Numerous documents and articles previously written by Dominique Paret and prepared for other technical publications have been included and adapted: [PAR 03, PAR 05] and [PAR 12].

Introduction

This book is specifically devoted to the design (theoretical and practical) of antennas for near-field communication (NFC) devices, irrespective of their environment, size and application (readers, tags, mobile phones in card emulation mode, peer-to-peer (P2P) mode, tablets, communicating objects, TVs, photo frames, etc.).

Before going into detail about the content of the individual chapters, we assume that readers already have some degree of familiarity with the field of radiofrequency (RF) and know roughly what the term "NFC" covers. Therefore, after a brief recap of the basics of the technical constraints and the normative framework which must always be borne in mind, we will carry out a detailed examination of not only the designs of 13.56 MHz antennas for NFC "*initiators*" and "*targets*", but also of all the problems of mutual interactions between initiators and targets, which must always be taken into account in order to ensure the correct operation of an NFC setup.

Also, readers should take note right now that when reading this book, they should always be conscious of the problems that may be caused by the physical layer (the medium and its management). The challenge here, then, is to present this book in as appropriate and clear a way as possible to help in understanding all these issues. After much reflection and discussion, a choice had to be made in favor of an overarching, clearly-didactic presentation, so that readers can easily orientate themselves in the maze of all these laws of communication, which many of our readers will discover and employ in years to come.

The division of the table of contents gives a very clear roadmap to follow in order to gain a helpful view of the concrete design of antennas for all sorts of NFC devices, and of the state-of-the-art in the field.

And now, dear friends, to work.

PART 1

Context

Introduction to Part 1

To transform any given system into a near-field communication (NFC) system, on paper, we only need an integrated circuit, some miniature passive components (R, L and C – resistance, inductance and capacitance) and an antenna. This is the theory. In reality, this migration is far more complex because, in view of the vast field of application of NFC, there are numerous constraints that need to be dealt with, respected and satisfied in order to render the final solution viable and reliable.

Although this first part of the book is an introduction, it encapsulates numerous technical points which are important to know in order to truly appreciate the applicational problems of today's world, as well as tomorrow's, specifically in terms of the antennas for these devices. Overall, it is an introductory recap: one which is absolutely crucial for readers to orientate themselves and to understand this book. It is divided into two chapters:

– Chapter 1 gives a brief overview of the constraints to which NFC systems and the connected problems that are directly associated therewith;

– Chapter 2 offers a brief recap of some fundamental physical laws governing the working of NFC.

It is now for you, the readers, to discover all this for yourselves.

1

Recap of the Constraints Governing the Design of Antennas for an NFC Device

In the interest of understanding, let us begin with a few terms of the vocabulary specific to the norms and/or standards of radio frequency identification (RFID), contactless and near-field communication (NFC).

Table 1.1 offers a few examples of jargon applied in different fields of application.

ISO group		Transmitter	Responder
		Base station	Target/transponder
		Reader	Card
		Modem	
		Coupler	Badge
SC 17 WG8	Proximity cards and personal devices	PCD Proximity coupler device	PICC Proximity integrated circuit card
	Vicinity cards and personal devices	VCD Vicinity coupler device	VICC Vicinity integrated circuit card
SC 31 WG 4	Item management/RFID	Interrogator	Tag
SC 06	NFC	Initiator	Target
		etc.	etc.

Table 1.1. *ISO terminology for the main contactless transmitters and responders*

In this book, which deals exclusively with NFC (originally developed by the ECMA in Switzerland, and then taken up again by Sub-Committee SC 06 at the ISO in 2000), we will employ only the official ISO terms: "*initiator*" and "*target*". Thus, from this point on, having looked briefly at the principle above, all other terms will be proscribed (or almost).

In the context of NFC applications, this chapter recaps the context, numerous constraints, functional and structural problems relating to the intrinsic content of the NFC protocols and those connected to it, and their direct implications in terms of antennas, which must be dealt with in order to be worthy – in the legal sense (to prevent lawsuits for false advertising) – of the label "Complies with NFC ISO 18092 or 21481 or NFC Forum standard" (in active or passive mode, batteryless or battery-assisted, etc.).

As we will see later on, there are various kinds of such issues.

1.1. Normative constraints

When designing an NFC system and the associated antennas, the technical and protocol constraints needing to be respected are, obviously, the legislational and physical constraints pertaining to the "low layers" 1 and 2 of the open systems interconnection (OSI) model (which are, respectively, the physical and data link/medium access layers), without which the whole setup could never work. As the antenna is part of layer 1 – the physical layer – for all intents and purposes, it is the center of the world.

The forms (appearances, amplitudes, etc.) of NFC signals that must be respected are described in detail in the *international standards* ISO 18092 NFC IP1 and ISO 21481 NFC IP2 – the lone true standards – which draw extensively on the contactless proximity chip-card standards ISO 14443 A & B (including numerous classes of antennas – 1-6) and on Japanese standard JIS X6319-4 for the patented product FeliCa, and finally those surrounding ISO 15693. In addition to these, there may be *proprietary- and/or market-sector-specific standards* such as (mainly) NFC Forum, EMVCo and CEN, where the specific operational application characteristics of distances and volumes (in cm^3) very frequently involve antennas with different adaptions.

To conclude this introduction, the normative framework of NFC aside, in this book the exchanges taking place between initiators and targets are defined, once and for all, as follows:

– "from the initiator to the target" known as *uplink*;

– "from the target to the initiator" known as *downlink*.

1.1.1. *Uplink from initiator to targets*

In order to avoid any comprehension problems, note now that, regardless of the intelligence built into the target, it only functions on the basis of "commands" sent by the initiator, which is a *TRANS*mitter. The initiator also includes a re*CEIVER* to pick up and interpret communication in the other direction.

Therefore, the initiator is a TRANS… CEIVER; a TRANSCEIVER.

Once again, in order to prevent numerous potential cases of confusion, two potential scenarios are officially defined by the ISO (ISO 19762-3 – Information technology – Automatic identification and data capture (AIDC) techniques – Harmonized vocabulary – Part 3: Radio frequency identification (RFID):

– either the energy transmitted by the radio frequency (RF) wave from the initiator is able to provide the necessary power to the target, and in this case the target is "*remote-powered*" or "*batteryless*";

– or the energy transmitted by the RF wave emitted from the initiator is unable to remotely power the target (which may be the case because of the desired distances of operation, the technologies used, the regulations in force, hostile environments, etc.), and of course, it is important to do something about that. In this case, the target is "*battery-assisted*", and we are dealing with other types of antennas.

NOTE.– Very frequently – too often, in fact, and *incorrectly* – remote-powered targets are said to be "passive", while battery-assisted targets are said to be "active". This makes absolutely no sense (see the explanations later on for details).

Nevertheless, it is impossible to put the point across more clearly. Here again, it is important to try to choose the right word.

1.1.2. *Downlink from targets to initiator*

Independently of the type of power supply to the target (remote-powered or battery-assisted), that device must have an electronic means of communication to perform a downlink from the target to the initiator – also known as the "return link". Downlink can take place in a number of different ways depending on the principles which are used.

It is important not to confuse the power/energy transfer system with the principles of upward and downward communication.

1.1.2.1. *"Passive" targets*

The adjective "passive" refers to the fact that the downlink – from the target to the initiator – takes place without the use of an RF transmitter.

1.1.2.2. *"Active" targets*

However, regardless of the way in which the target is powered, if it has a transmitter built in to respond to the initiator, it is said to be "active".

1.1.2.3. *Load modulation*

In order to achieve downlink, the initiator provides a physical support in the form of a sustained, non-modulated "carrier" frequency, and allows the target to act however it seems fit depending on its own way of working, in order to communicate with the initiator by modulating its electrical characteristics. For this purpose, there are two very similar modulation techniques, and one of which could be considered a "distant relative":

– the first, based on a principle of "modulation of impedance (resistance and/or reactance) of the target antenna's load" – known as *passive load modulation* (PLM) is used by most of the targets available on the market;

– the second, which is more recent, is known as *active load modulation* (ALM), and is still independent of the mode of power supply to the target (whether it is tele-powered or battery-assisted). The target is equipped with a (mini) low-power local transmitter so that just for that period of time, the signals returned to the initiator can be "boosted"..., i.e. sent with a newly adapted antenna.

In view of the physical consequences which arise from these two forms of near-field modulation, we typically speak of "retro-modulation" by "magnetic coupling".

In certain systems (such as NFC devices in "peer-to-peer" mode), during the downlink phase in a *half-duplex* system, the initiator no longer provides a carrier to serve as a support for the return signals. In this particular case, known as *active-mode NFC*, in order to communicate with the initiator, the target transmits its own wave, calibrated to the same frequency as that of the carrier, and becomes an "active" NFC device.

1.1.2.4. *Retro-modulation voltage*

Once it has sent interrogation commands, the initiator switches to *listen mode*, waiting for responses from the target. To this effect, with the exception of P2P mode, the initiator transmits a pure carrier consistently and waits, in the interests of comprehension, for the target to signal its presence and respond by a particular modulation which represents the variation of its load, whether that modulation is passive PLM or active ALM (see the previous sections 1.1.2.2 and 1.1.2.3).

1.1.2.5. *Retro-modulation voltage in PLM*

In reciprocity of the phenomenon of mutual induction between a secondary party (target) and a primary party (initiator), the value of the variation of voltage induced in the coil of the initiator's antenna during the return modulation is "ΔV1". Evidently, this variation in voltage is all the greater when there is a significant distance between the target and the initiator, because the coupling coefficient and the mutual are less.

When the target is situated very close to the initiator's antenna (an example is a mobile telephone, held to a point of sale (POS) payment reader), the coupling coefficient "k" can be up to 20 or 30%, and we need to take account of the presence of a *shunt* or a *loading effect*.

This voltage "ΔV1" is added to that which is already present at the edges of the initiator antenna. Hence, the resulting signal is the carrier, slightly modulated by that return signal, and with a consequent modulation index of a few percent. The signal which is present at the edges of the initiator antenna is also radiated, with its specific spectrum, and therefore its own side bands, which must, of course, fit into the Federal Communications Commission (FCC) and European Telecommunications Standards Institute (ETSI) patterns.

In conclusion, Table 1.2 sums up the passive, active, remote-powered, battery-assisted and communication functions.

"Supply" versus "tag to interrogator communication" (*Harmonized Vocabulary* ISO 19762 part 3)		
	Communication from tag to interrogator	
	Via load modulation	*Via transmitter*
Supply		
No battery on board	*Passive* *Battery less*	*Active* *Battery less*
Battery on board	*Passive* *Battery-assisted*	*Active* *Battery-assisted*

Table 1.2. *Passive, active, remote-powered and battery-assisted functions*

Now that this little review of the vocabulary has been carried out, we can focus on the purely technical aspects, examining how passive and active targets work.

1.1.3. *"Contactless" standards versus NFC device antennas*

Table 1.3 shows the normative consequences on antenna design of the main significant technical points of proximity contactless chip cards (ISO 14443 and JIS X6319-4) and those for vicinity contactless chip cards (ISO 15693), considered to be a "legacy of the existing technology", which were then joined by the additional points to create the concept and standards of NFC.

Once again, we need to look at a few key points of vocabulary specific to NFC, which have now emerged from the concatenation of the definitions given by ISO 18092 (NFC communication protocols IP 1 and 2 for the air interface) and the specifications of the NFC Forum and the design of their antennas.

	Chip cards			NFC			Immediate consequences for antenna design
	ISO	EMV	ISO	ISO		NFC Forum	
	proximity		vicinity	IP1	IP2		
	14443	EMVCo L1	15693	18092	21481	Analog/DP	
14443-2	yes	yes		yes	yes	yes	
							Q for proximity
Bit rate in kbps		106 only		106 and 212-424	106 and 212-424	106 to 424	Q for vicinity
							Bp for 106
							Bp for above
Field H in A/m	1.5 to 7.5	not specified	0.5 to 5	1.5 to 7.5	1.5 to 7.5	not specified	EMV & Forum = not specified
Retro-modulation							ISO = EMV & Forum another possibility
14443-3	Yes			yes	yes	yes	Target management, and thus stacks to be managed
Data collisions	yes	without anti-collision	yes	yes	yes	yes	Multi-card prohibited, and thus no stacks
RF collisions	no	No	no	yes	yes	yes	
Active mode	no	No	no	yes	yes	no	Dynamic initiator/target turnover
Tests	10373-6	EMVCo L1	10373-7			not available	
							ISO = field
Field H in A/m	yes	no	yes	yes	yes	no	EMV & Forum another possibility
Antenna class	yes (6)	1	not specified	not specified	not specified	yes (3)	
		3				3	
Volume	no	yes	no	no	no	yes	ISO = no volume imposed EMV & Forum two different volumes
Distance	no	yes	no	no	no	yes	ISO = no distance imposed EMV & Forum two different distances

Table 1.3. *Normative consequences for antenna design*

1.1.4. *Technologies*

Beyond the ISO NFC IP2 and IP1 standards, the NFC Forum defined three generic groups of transmission types under the term "technologies".

1.1.4.1. *NFC-A technology*

"NFC-A" technology is very similar to ISO 18092 NFC IP 1 and IP 2 which, themselves, were heavily inspired by the contactless chip card standard ISO 14443 type A, parts 2 and 3. In the NFC Forum, NFC-A is used for the tags called T1T, T2T, T4AT and P2P.

1.1.4.2. *NFC-B technology*

"NFC-B" technology is very similar to ISO 14443 type B, tirets 2-3, which had already been reused solely in ISO NFC IP 2. In the NFC Forum, NFC-B is used for the tags called T4BT and P2P.

1.1.4.3. *NFC-F technology*

"NFC-F" technology is very similar to ISO 18092 NFC IP 1 and IP 2 which, in large part, were heavily inspired by the Japanese standard JIS X6319-4. In the NFC Forum, NFC-F is used for the tags called T3T and P2P.

1.1.5. *"NFC Forum Devices" and "NFC Forum Tags"*

Even more new terminology.

1.1.5.1. *NFC device*

The ISO standards pertaining to NFC use this term to denote a communicative "initiator" or "target", be it in active mode or in passive mode (see section 1.1.6). For its part, after dividing up different classes of "technologies" (see section 1.1.4), the NFC Forum also decided to define and separate all "NFC devices" (from the ISO) into two classes of elements:

– first, *NFC Forum Devices*;

– second, *NFC Forum Tags*.

1.1.5.2. *NFC Forum Devices*

Whatever their physical forms (a mobile phone, a PC, a TV, a camera, a photo frame, tablets, etc.), *NFC Forum Devices* are elements which support the different modes and roles of operation described below, for all products and standards supported by the NFC Forum, which implement at least the *mandatory* parts of the protocol *stack* issued by the NFC Forum, and satisfy the demands and necessities of interoperability imposed by the NFC Forum.

No particular form factor is specified for NFC Forum Devices, but they:

– *must* (*mandatorily*) communicate with one another in accordance with ISO 18092, the protocols attached to the NFC Forum LLCP, "P2P", read (and write) the different types of NFC Forum Tag and be able to function as an initiator, target and reader/writer;

– *should* (*optionally*) offer the possibility of emulating a contactless card. Thus, an NFC Forum Device can be used as a contactless chip card of type ISO 14443 A (such as MiFare) or B or FeliCa (via the platforms T3T, T4AT or T4BT).

As we can see, an NFC Forum Device is a technologically complex element (and so is its antenna), because it is mandatory for it to:

– implement multiple technologies (A, B and F) on board;

– be able to switch from listen mode to poll mode;

– be able to discover whether there are other NFC Forum Devices or NFC Forum Tags in its vicinity, and process RF collisions and data collisions;

– communicate with those other devices.

1.1.5.3. *NFC Forum Tags*

Once again, no form factor is specified for these elements, but often a use case is envisaged: that of a *smart poster*, which essentially means nothing and tells us nothing about its form factor – is it round, square, small, large? It is a mystery. More formally, an NFC Forum Tag is a contactless tag or a (smart) card supporting the protocol *NFC Data Exchange Format* (NDEF) on a passive communication, and consequently, an NFC Forum Tag only transmits responses and supports at least one legacy communication protocol, defined in the specifications of the NFC Forum.

Let us now briefly look at the definitions of the "modes" and "roles" which NFC devices may take on during their respective operations.

1.1.6. *Modes of communication of an NFC Forum Device*

By dividing its time, an NFC Forum Device can support different "modes" of operation. There are four such modes which pertain to communication between elements. However, the word "mode" also covers two different, unrelated concepts.

To begin with, two suboptions of *modes of communication* are described in the standards NFC IP-1 (ISO 18092) and IP2 (ISO 21481).

1.1.6.1. *Passive communication mode*

In this mode, the NFC Device "Initiator" generates an RF field, which it modulates in order to send commands. The NFC devices "*Targets*" present in the field send their answers by load modulation (see Figure 1.1).

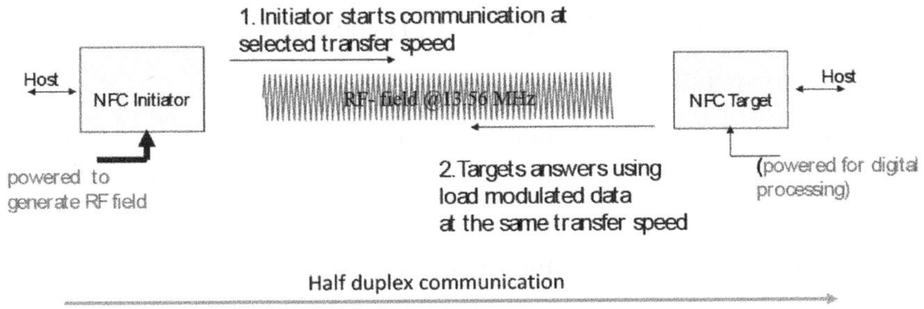

Figure 1.1. *Passive communication mode*

1.1.6.2. *Active communication mode*

In this mode, the NFC devices take it in turns to generate an RF field. The initiator begins by generating a field to send commands, and then stops totally. Then, in turn, the target produces its own field and modulates to send its answers (see Figure 1.2).

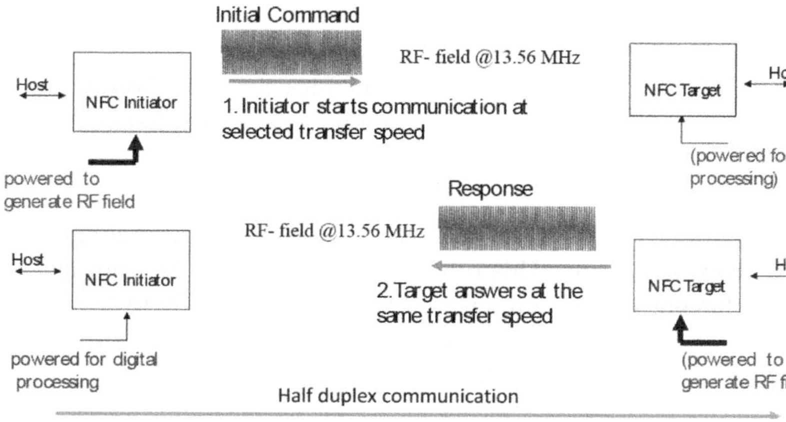

Figure 1.2. *Active communication mode*

Furthermore, two other modes – *listen mode* and *poll mode* – define how communication is established.

NOTE 1.1.– These communication modes are completely independent of how the NFC devices are powered. There is no relation with the notions of remote power supply (batteryless) or battery-assisted operation of the NFC devices.

NOTE 1.2.– Whether the NFC devices (the initiator or the target) are in active or passive mode, transport of the data in the two directions takes place in an alternating manner, rather than simultaneously. Thus, communication is *always half-duplex*. This remark remains applicable for the whole of this book.

NOTE 1.3.– Although many people use the term *peer-to-peer*, abbreviated as P2P, to speak of "active mode" the term *appears nowhere in the ISO standards*. Indeed, with no pun intended, "active mode" does not, in itself, constitute a true P2P model. A P2P model is a *client–server*-type model, in which each client is also a server, and this is not exactly the case because, in active mode, there is always the same server: the initiator which started off the exchange remains in charge of that exchange for the whole duration of the communication, and the client remains the same: a target.

1.1.7. *Role of an NFC Forum Device*

The NFC Forum defined four different roles for NFC Forum Devices, depending on the mode and communication protocol used.

1.1.7.1. *Initiator*

The "initiator" is the NFC device which generates the RF field and which begins the communication. This role is played by an NFC Forum Device in poll mode, having passed through a number of stages (*Activities*). In this role, the NFC Forum Device communicates by using the NFC-DEP Protocol (which is very similar to NFC IP1 – ISO 18092).

1.1.7.2. *Reader/writer*

This role is played by an NFC Forum Device in poll mode having passed through a certain number of activities. In this role, the NFC Forum Device acts as a "contactless reader" (in read/write mode), using a protocol and commands "inherited" as a legacy from one of the "Technology Subsets" (platform T1T, T2T, T3T or T4T).

The function of an initiator in poll mode, operating as a reader, is that of a device which reads, for example, simple tags to discover their content, but is unable to alter their content – for instance, we might read a poster on the street. However, note that: the term "reader" is always fallacious, because it often carries with it the implied double meaning of reading and writing, rather than that of reading "*only*". This being the case, the purpose of the NFC device is not only to read content, but also to be able to modify that content – for example, during a stock-take.

1.1.7.3. *Target*

An "NFC device" becomes a target when it responds to the initiator's commands, replying by way either of load modulation (with the RF field always being generated by the initiator) or modulating its own generated field. This is a role played by an NFC Forum Device having passed through a number of activities. In this role, the NFC Forum Device communicates by using the NFC-DEP Protocol.

Beware of the term *target*. In terms of functions, a target may be very powerful (a target in active mode – i.e. one which can also transmit), or a target may be in passive mode (and act as a simple tag), or indeed operate in card-emulation mode.

1.1.7.4. *Card emulator*

The role of a card emulator is achieved when an NFC Forum Device is in listen mode. The NFC Forum Device then behaves like one of the technology subsets and a legacy protocol (T3T or T4T platform). In this role, the NFC Forum Device communicates using the ISO-DEP Protocol.

1.1.8. *Beware of false advertising*

To conclude, and to state it plainly (to call a spade a spade), before proudly branding a product with the NFC label and declaring it "compliant with NFC standard ISO 18092 or 21481 or the NFC Forum standard" (of course, in active or passive mode, batteryless or battery-assisted, etc.) in the legal sense (otherwise, we risk lawsuits for false advertising), it must be at least worthy of that declaration, rather than being merely (though not pejoratively) a simple ISO 14443 system, as is often the case. This is a gross misuse of language.

1.2. Regulatory constraints

1.2.1. *RF regulations*

All the NFC systems described in this book have antennas, capable of transmission, arranged at the initiator and also at the target. There are very numerous regulatory texts which, by way of emitted frequency values, authorized radiation levels and specific patterns (templates), stipulate the constraints and restrictions (radiation, pollution, susceptibility, etc.) to which equipment designed for contactless applications, in the broad sense of the term (contactless chip cards, RFID, NFC, geolocalization, etc.) is subject.

At the frequency of 13.56 MHz used for NFC, the regulations and norms on radiation and pollution are essentially harmonized throughout the world, although there are some minimal variations from one region to another.

In the United States, the FCC – Federal Communications Commission – under the aegis of the ANSI (American National Standards Institute) established the famous reference document "US Code of Federal

Regulations (CFR), Title 47, Chapter I, Part 15, 'Radio Frequency Devices'", which is the local bible for regulations.

In Europe, the recommendation in force is the ERC – REC 70 03 "Relating to the use of Short Range Devices (SRD)", issued by the CEPT/ERO (European Regulation Organization, based in Denmark). Readers are strongly advised to obtain a copy of the latest version (available at www.ero.dk). In addition, the ETSI's measuring and testing methods follow the ERO recommendations: "Electromagnetic compatibility and Radio spectrum Matters (ERM) – Short range devices": EN 300 – 330 – frequencies from 9 kHz to 25 MHz, so applicable for 125 kHz and 13.56 MHz.

In Japan, the Association of Radio Industries And Businesses (ARIB) specifications are obeyed.

1.3. Constraints on the NFC market

If we create a product, it is because we hope to sell it. Thus, it needs to be "saleable" (and better yet, "buyable"); its function, usefulness, sale price at entry onto the market, the decrease in that price over time, profitability, etc., need to be known. That is to say, we need to carefully study the famous "Business Model" before beginning.

The NFC market is subdivided, overall, into two main domains: the markets for "niche" applications and the markets for "mass" applications. Often, NFC systems are intended for the general consumer market – i.e. for mass distribution – so the costs are a crucial factor. This, in one line, is the brief description of the techno-economic chain put in place to satisfy users' wishes and desires:

inexpensive → batteryless → remote-powered → radiated power within the authorized limits.

Inexpensive, and therefore with no battery on board the element. Here, the term "batteryless" means that the system's local power supply must come from a principle known as "remote power", using the near-HF field radiated by the initiator;

One of the problems readers may rightly ask is whether remote power supply is possible while respecting the RF regulations in force (bands, frequencies, powers and/or maximum authorized radiated fields H or E, patterns such as data rates, bit codes, collision management, cyclic usage report, etc.).

1.4. Typological constraints of NFC

No, no and no – the applicational typology of NFC is not linked only to mobile telephony. No, NFC is not merely a contactless payment card (emulator). No, because these views are highly reductives, and there are numerous applicational typologies of NFC, which we will now briefly review, along with the attendant consequences for the design of their antennas.

Table 1.4 summarizes the range of "reader" applications (RFID, P2P, battery-assisted passive card emulation, remote-powered passive card emulation, etc.) engendered by the physical fundaments of the NFC system.

	NFC device								
	Initiator		Target						
Means of communication		Role		Role					
	Initiator			Tag	Card		Down-link		
Passive Half-duplex Bidirectional	P	Reader/writer	L	T1T, T2T, T3T or T4T poster			PLM *ALM*	Batteryless	Poster, label
	P	Reader/writer	L		T3T or T4T chip card		PLM *ALM*	Batteryless	Ordinary chipcard. Bank card, loyalty card
		Reader/writer	L			T3T or T4T cardemul.	PLM *ALM*	Batteryless	Phone with flat battery, loading effect
						T3T or T4T card emul.	PLM *ALM*	Battery-assisted	Normal phone, loading effect
							ALM not described in NFC transmits	Battery-assisted	
Active Half-duplex Bidirectional	P	Initiator NFC-DEP Protocol NFC IP1 – ISO 18092			Other device			Battery-assisted	

Table 1.4. *Range of NFC applications*

1.4.1. *Application consequences and their direct constraints*

From these generic possibilities derive applicational services, which are generally grouped into the following for main marketing terms: "*Touch & Go*", "*Touch & Confirm*", "*Touch & Connect*" and "*Touch & Explore*". In parallel, the applicational diversity in terms of the architectures and electronic functions of NFC devices is very great, and the "ecosystem" associated with NFC is very difficult to establish, as the tie-ins, overlaps and interapplications are so numerous, both at economic level and at the level of technical solutions to be put in place and resolved.

Table 1.5 gives numerous examples of the applicational technical diversity of NFC and the forms, dimensions and form factors of antennas toward which we are advancing (… with no attempt at publicity).

		Architectures				
		Initiator	Target		Non-exhaustive examples	
Markets	Functions	Writer/ reader	Tag	Chip-card emulation	Dimensions of the antennas (order of magnitude)	Non-exhaustive examples of manufacturers
Loyalty card	Badge chip cards		x x		5 × 6 cm	
Mobile telephony	Smart phones All applications	x		x	6 × 6 cm to 3 × 3 cm	Google, Apple, Samsung, etc.
	Add-on cards			x	2.5 × 2.5 cm	Mercury
Banking payment	Smart card			x	5 × 7 cm 2.5 × 7 cm	Gemalto, G&D Oberthur, etc.
	MPOS – *Mobile Point of Sales*	x			5 × 4 cm screen circumference	Ingenico, Wincor Nixdorf, etc.
	Tablets, wallets	x	X	x	above and below	
	DAB player	x			Φ = 7 cm EMVCo	Thales, etc.
Industrial	Badges		X		Any shape and size!	
	Tags		X			
	Smart poster		X			
	Racing numbers		X			Dag System etc.
Medical	Identification, monitoring	x	X		4 × 1 cm	Trixell, Maquet
	Decontamination bin	x			30 × 30 cm	IMS
	Monitoring of babies, mice, ants, bees	x	X		Very small 4 × 1 mm	Lutronic, Arelis, Nonatec, etc,
Consumer	Photo frames		X		30 × 20 cm Screen circumference	Parrot, etc.
	Acoustic chambers		X			Parrot, etc.
	TVs	x		x	Secure access	Philips, etc.
	PCs	x		x		Dell, etc.
	Advertising items	x	X		All formats	Strapmedia
	Museum guides		X		Φ = 2.5 cm	Strapmedia
	Wireless chargers, libraries	x			10 × 8 cm	Tagsys, etc.
Computer technology	USB dongles	x	X	x	1 × 1.5 cm	Mercury, Neowave
	USB keys		X		2.5 × 2.5 cm	Neowave

	Terminals	x				Taztag
Automobile	Car dashboards					Herman
	Contact keys		X	x	3 × 5 cm	Conti, Valeo
	NFC accu. charger.	x			10 × 8 cm	Delphi Conti
Public transport	Access validator				Φ = 10 cm	Xerox, Parkeon Thales, etc.
	Control device				5 × 3 cm	
IoT Connected "things"	Watches, keys				2.5 × 2.5 cm Φ = 2.5 cm	Apple Mercury
	Bracelets				1.5 × 3 cm	Apple, Mercury
	Phone booths					
	Social networks Community					Zèbre
Luxury items, jewelry	UHF and NFC in HF					
	Lingerie, glasses, shoes, etc.		X		Combatting counterfeiting All formats	Louboutin, Lise Charmel, Ray-Ban, etc.
	Wines, spirits, LCD labels, corks		X			
			X			
	Jewelry store windows				Security	

Table 1.5. *Examples of the diverse technical applications of NFC*

As readers may realize, the mechanical ingenuity of the marketing department, designers and developers of all these markets is limitless. However, in terms of magnetic fields and communication in the surrounding environment (passive or active communication, reader/writer mode, P2P, emulation card, loading effect, etc.), let us simply say, diplomatically, that it is a different story.

In addition, to serve the needs of applications in banking, transport, etc., we often need to bring the initiators and targets operating in contactless chip card emulation mode and the performances of the magnetic fields generated by their antennas back within the limitations of the technical specifications of "EMVCo Contactless", which are particular implementation specifications reviewed and adapted from ISO 14443 A & B, which are encapsulated by the NFC IP2 standard. Thereafter, of course, we need to run all the infamous tests on common criteria, and obtain all the labels such as EAL 5+, 6+ and even better.

1.5. Applicational constraints on antenna design

Whether an NFC device is an initiator or a target, in active or passive mode, batteryless or battery-assisted, the very numerous applicational constraints and their direct relations with the design of an antenna arise primarily at seven or eight major levels, presented in the form of three distinct domains to clearly define each one's role.

To begin with, the *structural problems* linked to the direct and indirect consequences of the intrinsic content of the NFC norms and standards (ISO, NFC Forum, EMV, etc.). These include:

– "*form factors*" of the antennas;

– the variations in operational distance, and therefore of magnetic couplings;

– the values of coupling coefficients and indicators;

– the values of retro-modulation voltages;

– the values of quality coefficients;

– the various modes of operation per slice of time.

Then, the *functional applicational problems* pertaining to the different types and the variability of applications (loading effects on the targets, stack effects, etc.) dealing with:

– initiators and targets that are deliberately not synchronized and/or have accidentally become desynchronized;

– problems relating to "stacks" of NFC devices.

Finally, we face the *environmental problems* arising directly from the antenna and how to get around them either partly or completely. These include:

– environments hostile to wave propagation;

– the effects of boxes, metal parts, the presence of the battery;

– electromagnetic compatibility (EMC) pollution.

Readers can also refer to an entire book devoted to the constraints involved in the design of an NFC device: [PAR 16], from the same publisher.

2

Introduction to and Recap of the Principles Employed in NFC

In order to be able to look at the idea of near-field communication (NFC) in its proper context, this chapter offers a mini-recap of the main physical principles used in NFC, which must always be borne in mind throughout this book.

2.1. The physical fundaments of "contactless" and NFC

To aid in the understanding of the specific problems of these applications, for readers who are unfamiliar with the technical details (or who need to refresh their memories), let us give a brief reminder here about radiofrequency (RF) communications, and describe unambiguously what the term *near field communication* covers. It is physics; not marketing. Hence, it is necessary to review some of the theoretical and physical bases of the technical side of NFC.

2.1.1. *Phenomenon of propagation and radiation*

Let us look at some of the conclusions on the physical concept of propagation of an RF electromagnetic (EM) wave in the classic case of radiation of a simple "Hertzian doublet". Later on, we will use this simple example to aid an examination of concrete cases of NFC.

2.1.2. *Classification of fields and spatial regions*

When studying the phenomenon of propagation and radiation of EM waves, the Maxwell equations tell us that the values of the radiated electric field E and magnetic field strength H depend notably on three main parameters:

– the variable time t;

– the variable distance r from the radiation source, defined in terms of $1/r$, $1/r^2$ and $1/r^3$. If one or more of these terms is negligible in comparison to the other(s), the equations may be simplified;

– the product kr, equal to: $kr = 2\pi/\lambda$.

2.1.3. *Spatial regions*

In the case of a very small antenna (whose sphere equivalent to the radiation source has a very small diameter), it is customary to define three spatial regions known, respectively, as the "near-field, intermediary-field and far-field zones" depending on the possible values of r in relation to the value of the wavelength λ at which the system operates – related to the value of r is greater or lesser than:

$$\frac{1}{k} = \frac{\lambda}{2\pi}$$

2.1.4. *Far field: r >> λ/2π (Fraunhofer zone)*

In the far field zone, the terms in $1/r^2$ and $1/r^3$ present in the Maxwell equations become negligible, and the values of the fields E and H shrink in $1/r$. This occurs when the distance r is greater than a few $\lambda/2\pi$. What happens at this point is that we move from a region of quasi-stationary waves to a region of wave propagation, and thus to a regime of radiated field. The electrical field E is predominant, and there can no longer be any magnetic coupling. Generally, in this zone, devices work by using the radiating field E, so use a dipole antenna.

2.1.5. *Intermediary field: r approximately equal to λ (Fresnel zone)*

In this spatial region, all the terms in $1/r$, $1/r^2$ and $1/r^3$ are equally preserved. Here, we are in the presence of a radiating near field, where the Fresnel component exists when the antenna's primary dimension is larger than the transmission wavelength λ but the distance r is approximately equal to λ.

2.1.6. *Near field: $r \ll \lambda/2\pi$ (Rayleigh zone) ... and by essence, the origin of the "NF – Near Field", and hence NFC*

In the spatial region where $r \ll \lambda/2\pi$ – *irrespective of the frequency at which the system operates* – i.e. often in the immediate – or near surroundings of the antenna, the values of the terms in $1/r^2$ or $1/r^3$ are far greater than those of the terms in $1/r$, and the fields (E, H) decrease very quickly. This near-field zone – often called the Rayleigh zone – essentially corresponds to the region of a "(more or less) couple transformer" (where the received power P_r is basically equal to the transmitted power P_t, unlike the effect of free space propagation, where $P_r \ll P_t$ (far field)).

When the transmission takes place very near to the antenna, the electrical field is generated by the antenna's own potential: this is the near-field domain. At this distance, we see a regime of "quasi-stationary" waves, and it is then possible to establish magnetic – or electrical coupling between a tag and the source.

Generally, in this zone, devices work by using the magnetic field H by inductive coupling, so use a loop antenna, which is the case with NFC.

IN SUMMARY.–

$$\text{Near field (magnetic coupling)} < \frac{\lambda}{2\pi} < \text{Far field (wave propagation)}$$

2.1.7. *Remarks on contactless, RFID and NFC application*

Having made these general remarks, which are of fundamental importance, at a given date, for remote-powered systems (when the transmitted wave provides enough energy for the tag to function), for existing semi-conductor technologies and the regulations in force, as a preliminary point, it is always interesting to know the regions of the near-field and far-field zones which the contactless/RFID/NFC application envisaged is intended to function. With this in mind, let us look at Table 2.1, which indicates the relations between the values of the frequencies and the distances λ/2π associated therewith.

	Frequencies	Wavelengths	$\frac{\lambda}{2\pi}$	Near-field zone for NFC application	Start of far-field region >~3 λ
LF	125 kHz	2,400 m	382 m	Always	1,100 m
HF	13.56 MHz	22.12 m	3.52 m	Always	11 m
UHF	433 MHz	69.2 cm	11.02 cm	0 to ~12 cm	24 cm
	866 MHz	34.6 cm	5.51 cm	0 to ~ 6 cm	18 cm
	915 MHz	32.8 cm	5.22 cm	0 to ~ 5,5 cm	17 cm
	2.450 GHz	12.2 cm	1.94 cm	0 to ~ 2 cm	6 cm
SHF	5.890 GHz	5.1 cm	0.8 cm	0 to ~ 1 cm	3 cm

Table 2.1. *Relations between the values of frequencies and the distances λ/2π associated therewith*

As this table illustrates, in ultra high frequency (UHF) and super high frequency (SHF), most of the time the operating distances required/possible for RFID applications (for example, reading of labels on pallets of goods) are significant (several meters). Given that these distances are always much greater than the value of λ/2π, the UHF tags used for RFID work in the "far field", and therefore must use principles other than that of magnetic coupling to communicate with the base station – for instance, they could exploit the phenomenon of propagation of EM types and backscattering of the received waves.

However, in the low frequency (LF) and high frequency (HF) bands, the operational distances required/possible for the applications – for technological reasons, because of the regulations in force, because of the possibilities of remote powering and because of cost factors – are always much smaller than the value of $\lambda/2\pi$. In this case, the devices operate and communicate in the "near field".

In conclusion, since their very beginning, in LF and HF applications (particularly at 13.56 MHz), contactless chip cards and RFID tags communicating at 125 kHz and/or 13.56 MHz operate in the "near field", and therefore, to paraphrase Monsieur Jourdain in *The Bourgeois Gentleman*, have been operating NFC all their lives and did not even know it. Thus, in terms of its principle, in LF and HF, NFC is not intrinsically any great leap forward. Hence, the applications are based on principles which primarily involve inductive loops and inductive/magnetic coupling.

2.2. The concept of NFC

Let us now examine the physics of the applicational concept of NFC.

Around 1999, the two main leaders in the semi-conductor market – Philips Semiconductors (a Dutch company, which has since become NXP) after having bought out the Austrian company Mikron; and Japan's Sony, the respective creators of the patented designs for contactless chip cards MiFare™ (based on the international standard ISO 14443 part A) and FeliCa™ (conforming to the Japanese standard JIS X6319-4) – came together to co-invent the concept of NFC for multiple, generic and varied applications for digital data, but not especially to supplant or compete with chip-card applications, as many people believe. Hence, NFC was born in around 2000.

To conclude this brief introduction, if readers still believe that NFC is synonymous only with mobile telephony, they are wrong, as proved by the following text from "Scope" (late 2002), as an introduction to the standard ECMA 340 NFC IP1, which was too become ISO 18092:

> "Near Field Communication Interface and Protocol (NFCIP-1) defines communication modes using inductive coupled devices operating at the center frequency of 13.56 MHz for interconnection of computer peripherals. It also defines both the Active and the Passive communication modes of Near Field

Communication Interface and Protocol (NFCIP-1) to realize a Communication network using NFC devices for networked products and also for consumer equipment…"

2.2.1. *Biot–Savart law*

Having clearly defined NFC and demonstrated its position in relation to other technologies, let us now give a recap of the physical theoretical minimum of what needs to be borne in mind in order to gain a detailed view of what is flowing through the coils and loops of antennas and what we are talking about throughout this book, because we must remember that an NFC device only works if layer 1 – the *physical layer* in the OSI model – works properly. This involves:

– the creation of a magnetic field H;

– the evolution of that field with distance.

The law, known as the Biot–Laplace and Biot–Savart law, establishes the mathematical relation which exists between a current I circulating in an element of an electric circuit of length dl and the intensity of the magnetic induction B as a function of the distance x existing between the measured point (of unit vector u) and the oriented element dl producing the field, according to the formula:

$$dH = \frac{(I\,dl) \wedge u}{4\,\pi\,x^2}$$

so that $B = \mu \times H = \dfrac{(\mu\,I)}{4\,\pi} \oint \dfrac{dl \wedge u}{dx^2}$

where $\oint$ is the curvilinear integral calculated on the length l and Λ is the notation of the vector product.

2.2.2. *Field H at a point on the axis of a circular antenna*

In the particular case of a flat circular antenna, with radius r, at a point P situated at a distance d along the axis of the antenna (Figure 2.1):

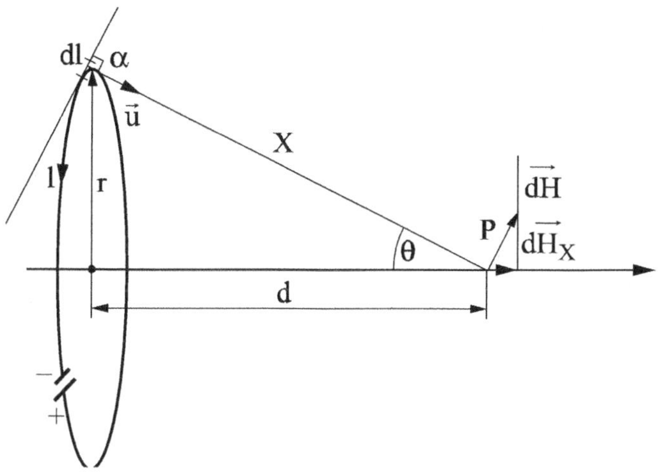

Figure 2.1. *Field H at a point on the axis of a circular antenna*

We can write:

$$\rightarrow dH = (I\ dl\ ) \wedge u)/4\pi\ x^2$$

and hence:

$$dH = (I\ dl\ \sin\alpha)/4\pi\ x^2$$

As, in this case:

$$\alpha = \pi/2 \text{ and } \sin\alpha = 1$$

so:

$$dH = I\ dl/4\pi\ x^2$$

On the axis of the coil, the component dH_x of dH is equal to:

$$dH_x = I\ dl\ \sin\theta/4\pi\ x^2 \text{ and } x = r/\sin\theta$$

and hence:

$$dH_x = I\ \sin^3\theta\ dl/4\pi\ r^2$$

In addition, at a distance d along the axis of the coil:

$$x = \sqrt{(r^2 + d^2)} \text{ and } \sin^3 \theta = r^3/x^3$$

so:

$$\sin^3 \theta = r^3/(r2 + d2)^{3/2}$$

which gives us:

$$dH_x = [\ (I/4\pi) \times (r/(r^2 + d^2)^{3/2})\ ]\ \times dl$$

Thus, definitively, by now integrating that simple integral in *dl* along the whole length of the circular spire of length $2\pi r$:

$$H_x = [\ (I/4\pi) \text{ x } (r/(r^2 + d^2)^{3/2}\ ]\ \int_0^{2\pi r} dl$$

we obtain:

$$H(d, r) = \frac{I\, r^2}{2(r^2 + d^2)^{3/2}}$$

Furthermore, $B = \mu\ H(d, r)$:

$$B(d, r) = \mu \text{ x } \frac{r^2}{2\ [(r^2 + d^2)\text{^}3/2]} I$$

EXAMPLE 2.1.– In the air, with $\mu = \mu\_0 = 4 \times 3.14 \times 10\text{^-}7$:

– if $H = 1$ A/m then $B = \mu\_0 \times H = 1.256\ \mu T$;

– if $B = 1\ \mu T$ then $H = B/\mu\_0 = 0.796$ A/m.

It should be noted that for d = 0 and a coil of N spires, we see the same, well-known formula for the induction B in the center of a flat circular coil:

$$B(0, r) = \mu \frac{N\ I}{2\ r} = \mu\ H(0, r)$$

2.2.3. *Decrease in the field H as a function of "d"*

By setting:

– *d* as the distance between the center of the antenna and the measuring point on the axis;

– *r* as the radius of the antenna;

– a = d/r and d = (a r).

The above relation between H_0 (in the center of the base station antenna) and H_d becomes:

$$H(a, r) = \frac{1}{[(1 + a^2)^{\wedge}3/2]} \times H(0, r)$$

The ratio H(a, r)/H(0, r) depending on the values of *r* and *d*, or indeed on the values of the ratio *a* = *d/r*, are shown in Table 2.2 (for a given value of r, the table indicates the values of the corrective factor "(1+a²)^3/2" and its inverse as a function of the ratio *a* = *d/r*).

d	a	H(0,r)/H(a, r)= (1 + a²)^3/2	H(a, r)/H(0,r)	= 1/(1 + a²)^3/2
0	0	√ 1	= 1	= 1
0.5 r	0.5	√ 1.953	= 1.397	= 0.716
r	1	√ 8	= 2.828	= 0.354
1.33 r	1.33	√ 21.34	= 4.62	= 0.216
1.414 r	1.414	√ 27	= 5.196	= 0.192
1.5 r	1.5	√ 34.33	= 5.86	= 0.170
1.66 r	1.66	√ 53.8	= 7.34	= 0.136
2 r	2	√ 125	= 11.8	= 0.0847
3 r	3	√ 1000	= 31.62	= 0.0316
4 r	4	√ 4913	= 70	= 0.0143
4.5 r	4.5	√ 9595	= 98	= 0.0101
5 r	5	√ 17576	= 132	= 0.0076

Table 2.2. *Variations of the ratio H(a, r)/H(0, r) depending on the values of "r" and "d"*

NOTE.– Throughout this book, we will very frequently refer back to this table, because it would be enormously problematic to repeat it whenever needed.

Figure 2.2 reflects the results given in this table.

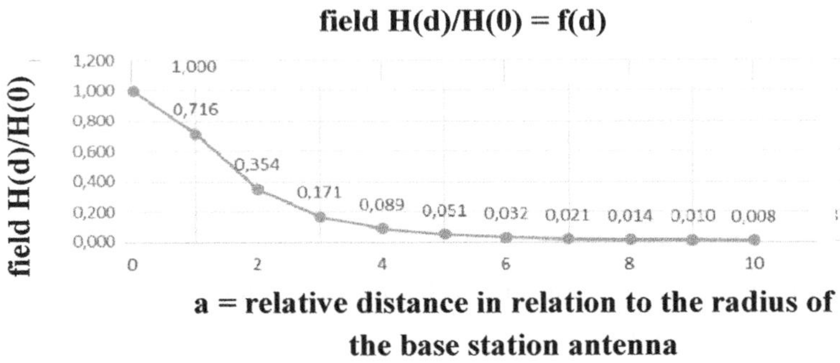

Figure 2.2. *Curve of the variations in the ratio H(d)/H(0) at "a"*

EXAMPLE 2.2.– For an antenna of an initiator/reader (proximity coupler device (PCD)) ISO10 737-6, so of radius r = 7.5 cm (diameter = 15 cm) we switch from a 7.5 A/m field at 0 cm to a 1.5 A/m field (i.e. a ratio of 5), when the ratio of the distance between the antenna of the PCD (reader) and the proximity integrated circuit card (PICC) (target) is around (1.4 × the radius of the PCD), which is a distance of 1.4 × 7.5 = 10.5 cm. On principle, a PCD yielding a magnetic field H(0) of 7.5 A/m can read all ISO PICCs (because it needs to be sensitive to at least 1.5 A/m minimum) from 0 cm to around 10.5 cm.

2.2.4. *Field H at a point on the axis of a rectangular antenna*

Similarly, in the case of a flat, rectangular antenna with sides a and b, with N spires, at a point M situated at a distance x on the axis of the antenna (Figure 2.3).

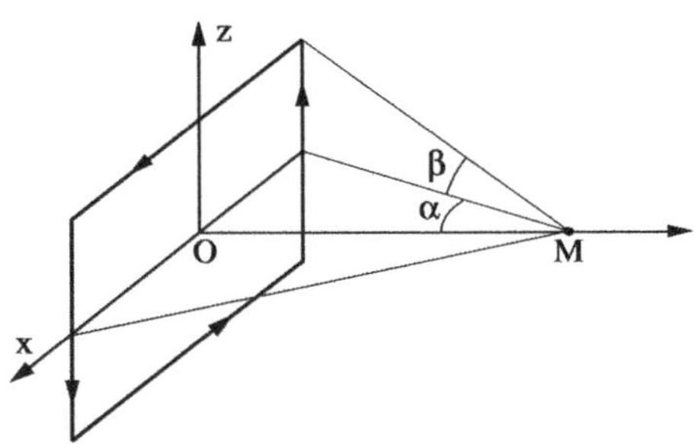

Figure 2.3. *Field H at a point on the axis of a rectangular antenna*

After a few lengthy calculations, we obtain:

$$H=\frac{N\times I\times a\times b}{4\times\pi\times\sqrt{\left(\frac{a}{2}\right)^2+\left(\frac{b}{2}\right)^2+x^2}}\times\left(\frac{1}{\left(\frac{a}{2}\right)^2+x^2}+\frac{1}{\left(\frac{b}{2}\right)^2+x^2}\right)$$

After these few physico-theoretical episodes, let us now examine the detailed calculations of NFC initiator and target antennas.

PART 2

Methods and Designs for NFC Device Antennas

Introduction to Part 2

The devices which are known internationally as *short range devices* operate either in *far fields* (at UHF from 860 to 960 MHz and 2.45 GHz, and at SHF at 5.8 GHz) by wave propagation (Maxwell equations), or in *near fields* (at LF around 125 kHz and at HF mainly at 13.56 MHz) by inductive coupling (Biot–Savart law). In the former case (UHF and SHF), we are interested in the electrical component E of the field, and the antennas belong to the "dipole" family; in the latter (LF and HF), which is our focus in this book, we look at the magnetic component H of the field, and the antennas belong to the "loop" family.

We now come to the most important part of this book. In this next part, we will describe many of the mysteries surrounding the computation – and encryption methods used for the physical layer – the "air interface" – which (too) many people allow to go unnoticed. Unfortunately, though, if the air interface does not work, or does not work well, nothing will. Therefore, it is preferable to ensure that it is firmly cemented.

For readers' comfort, this part of the book is divided into several chapters.

Chapters 3 (theories) and 4 (detailed examples) pertain to the computation methods and designs for "initiator antennas" for all form factors of initiator NFC devices, whose main function is to create and transmit the carrier signal at 13.56 MHz, in order to communicate with targets either in passive or active mode, whether they are batteryless (remote-powered) – in which case the initiator must supply energy (chip cards, ordinary NFC tags) – or battery-assisted (for example, a mobile phone in card-emulation mode), which are present within their fields.

Chapters 5 (theories) and 6 (detailed examples) "target antennas" look at antennas for targets in all forms:

– antennas for NFC tags;

– antennas for passive targets (with retro-modulation), e.g. functioning as card emulators;

– antennas for active targets (with retro-modulation), e.g. functioning as card emulators.

Before going into detail about this second part of the book, let us begin with some important preliminary remarks – two of them:

1) The contents of this part are not new. All the diagrams and couplings of antennas, and the calculations pertaining to them, which will be presented in the coming chapters, have been in existence since the very beginning of industrial activity in RFID, operating by magnetic coupling (near-field) – in concrete terms, since around 1985 in LF (125 kHz), with the component manufacturers NXP (Philips) Semiconductors, TI Texas Instruments and EM Microelectronics. A few others began introducing immobilizer applications and anti-starter systems for cars (with Siemens Auto/Continental, Valéo, Delphi, Sagem, etc.). Then, around 1995, Europe saw the introduction of magnetic coupling in HF at 13.56 MHz and of contactless payment cards, primarily by the Austrian company Mikron (which was later became Philips Semiconductors, and then NXP), and its flagship product MiFare. In Japan, there was Sony with its product FeliCa, and then finally, since around the year 2000, the teams of engineers at NXP in Graz (Austria) and Caen (France) have developed a broad range of NFC integrated circuits. Ultimately, suffice it to say that these are not new developments.

Today, these "theoretical and academic" diagrams and calculations are well-known, perfectly conventional bases; on occasion, some have been published by Philips/NXP in the form of application notes (ANs), and usually parsimoniously provided to specific users/customers, accompanied by Non-Disclosure Agreements (NDAs) and then happily reproduced a few years later by many other suppliers of components to the market, such as TI, EM, Infinéon, STm, Inside, Melexis, etc.

2) In this book, in turn, I have reproduced some of these classic and generic diagrams and calculations (after having undertaken original calculations both personally and professionally, 20 years ago, while employed by Philips/NXP), and added detail and further information here, because this book is intended to pass down technical and technological skills. It is worth noting that readers may well encounter the same kinds of technical problems as raised by the use of RFID/NFC communication devices, in developing new applications. Having read this chapter, readers should be able to develop the output stage for the RF part of their own RFID/NFC products without enlisting anyone else's help.

NOTE.– In this book, we will not present "exotic" antenna solutions, which are to be found in numerous patents, but whose applications are very specific, and in reality are rarely employed.

Acknowledgments

In terms of these next four chapters, I would like to offer special thanks to a number of long-standing friends, for their technical help with reflections, with content, with comments on the design of this long part concerning initiator – and target antennas and their combinations. These people are mainly Mme Catherine Maréchal, a researcher at the *Ecole supérieure d'ingénieurs* ESIGETEL, Jean-Luc Meridiano and Christophe Arnoux at Gemalto, Cyril Proye at Linxens, Denis Pietersoon, Johann Dantant, Jérôme Chalbot at ProActive, Serge Gasnier, Jean-Yves Cadorel and Eric Devoyon at CRESITT, Robert Chen and Fréderic Lhuillery at Parrot, Bernard Leibovici and Wilfrid Husson at IMS, Sacha Matic at IPDiA, and a great many more, along with the teams they lead. Many thanks to one and all.

3

"Initiator" Antennas: Detailed Calculations

3.1. Introduction

By definition, "initiator" mode or function in a near-field communication (NFC) device underlies the fact of establishing a communication with a target, and thus having a transmission function to transmit a wave in order to send a command, as well as a reception function in order to receive and listen for an answer from the target. In the initiator, these two functions may be totally separate, and have separate respective antennas – this is known as a "bistatic" solution – or else the transmission – and reception antennas can be combined into a mono-antenna – a so-called "monostatic" solution. For financial reasons (and of course simply because it works), this latter solution is chosen and employed in 97% of the market. Thus, it is this solution of "combined transmission and reception mono-antenna" which we will describe at length in this chapter on "initiator antennas"; a few lines at the end of the chapter will be devoted to bistatic systems.

3.1.1. *There are initiators ... and there are initiators*

Putting the antennas' physical size aside, let us begin with a detailed breakdown of the different types of "initiators" which underlie all the international norms (ISO and CEN), the proprietary standards (NFC Forum, EMVCo, etc.) and other terms commonly found in the press. In order to do so, we need to carefully divide these numerous possible initiators according to the three main parameters: modes of operation, principles of power supply and the types of operation of return channels (e.g. retro-modulation) which, for each case, imply antennas with different electrical designs, constructed from the same generic pattern.

Table 3.1 shows that breakdown.

	NFC mode of communication of the initiator		Initiator power supply		Retro-modulation from target to initiator		Examples of initiator applications in "reader/writer" mode
NFC device "initiator"	*Passive*	*Active*	*Battery-less*	*Battery assisted*	*Passive*	*Active*	
Ordinary "initiator" Mono- or bistatic antenna		x	.	x x	x		The initiator works in reader/writer mode and is capable of reading targets that are tuned or detuned, operating in PLM For instance: reader of chip card badges, "loyalty" on Posters, mouse, etc.
		x				X	As above, but the target operates in ALM
"Initiator" mobile phone mPOS payment		x		x	x	x	The initiator works in initiator reader/writer mode and is capable of reading targets that are tuned or detuned, operating in PLM or ALM For instance: Chip card payment reader in a mobile phone
	x			x	x		...As above, but, for each slice of time, the initiator reverses its function and also operates as a "card emulator", and can either by tuned or detuned For instance: PLM target chip card emulation for payment in a mobile phone
						X	ALM chip card emulation for payment in a mobile phone
"Initiator" *Peer to peer*		x		x	Active mode Transmission in each direction in turn		In P2P mode, there is an exchange between an initiator and a target. Hence, it is transparent for the load effect and any detuning

Table 3.1. *Possible points to satisfy the performance requirements*

3.2. Design of an initiator antenna (without influence from the outside environment)

By way of the numerous and varied examples discussed in Chapter 4, which are the fruit of lengthy experience in the field, the first paragraphs of this chapter present the breakdown of the different operations and stages, both theoretical and practical, leading to the complete definition and creation of antennas for NFC initiators. These different stages will be discussed one by one, but before we begin, in order to evaluate the radio frequency (RF) and antenna part of an NFC project, we have to describe its final goal, which is to *define/determine/discover the value of a current I flowing through an antenna (in our case, an inductance L) comprising N turns, in order to:*

1) first, create a magnetic field H proportional to $N\,I$;

2) second, engender a magnetic flux $\Phi = L\,I = B\,S = \mu H\,S$;

3) third, to produce an induced voltage $v = d\Phi/dt$.

In regard to the current I, it circulates in the antenna circuit which, in any case, will physically be a coil, so an inductance L, but will have an ohmic resistance *R_ant* and other losses, and thus, sooner or later, we will need to deal with a power level and losses in watts.

In addition, this setup will always be driven by a voltage generator with an internal ohmic impedance of source *R_gene_out* (an integrated circuit). The aim, then, will be to transfer the maximum watt power from the generator to watt load represented by the antenna circuit.

Alas, the two resistors (watt) *R_ant* and *R_gene_out* mentioned above (for numerous good reasons) have different values (Murphy and siblings) and, as we will now show, to optimize the maximum power transferred, we need to attempt, first, to adapt the impedance solely using purely reactive elements L, C, and second to tune them to preserve only the resistive term of impedance Z.

In addition, in parallel to this, as usual we need to design and satisfy filtering functions (to filter out harmonics, electromagnetic compatibility (EMC), etc.) – of course, with reactive, tuned loads (same causes and same effects).

Hence, the first part of this chapter will comprise long series of impedance matching from stage to stage, from A to Z.

3.2.1. *Operating mode*

Table 3.2 shows the breakdown of the different operations and phases which are necessary to perform in order to define the electrical and technological characteristics and the performances of a "loop" RF antenna for near fields, with inductive coupling. This is true *no matter what its form factors*, such as those used conventionally for a battery-assisted NFC initiator, whether it be a card reader, a terminal plugged into the mains, a mobile phone powered by a stack or battery in reader mode, or anything else.

Phase 0 – Brief instructive recap Phase 1 – We choose the type of integrated circuit for the initiator – We carefully read its data sheet – From it, we deduce the value of R_ic_out of the integrated circuit Phase 2 – We look at the pattern mask ERC 70 03 – We set the value of the filtering inductance, so that the capacity of the EMC filter is physically feasible Phase 3 – We impose the condition that the EMC circuit must be tuned to the carrier frequency – To transfer the maximum amount of power, we decide that Rs = R_out_ic – This gives us the values for Q and Q^2 – We find the value of Rs in relation to the parallel value Rp Then: Phase 4 – We look at the minimum operational field which the target manufacturer can guarantee – We define the distance at which we want to operate – We calculate the field that the coil of the initiator must provide at its center Phase 5 – With the above values, we determine the value of L = f(N) – We define the value L of the usable inductance, and thus its number of turns N Phase 6 – We define the value Rs_ant of the antenna in order to find the quality coefficient of the antenna Q appropriate for the application – We make the "overall antenna load", expressed in relation to its input terminals, a real impedance, equal to Rp via the values of the T-bridge matching circuit. Phase 7 – We lay the necessary cable, solder, measure and conclude.

Table 3.2. *Approach to designing the initiator antenna*

3.2.2. *Instructive recap*

3.2.2.1. *Conditions of power matching of a source/generator*

Let us give a brief recap so as to provide a proper understanding of our subject.

When we wish to *match the power* between the internal impedance $\underline{Z}_g$ of a sine wave generator and its load Z, certain conditions need to be satisfied. With this in mind, using the Thévenin equivalent diagram of the generator, we calculate the value of the load which needs to be plugged into its output in order for it to transmit the maximum amount of active power to its load.

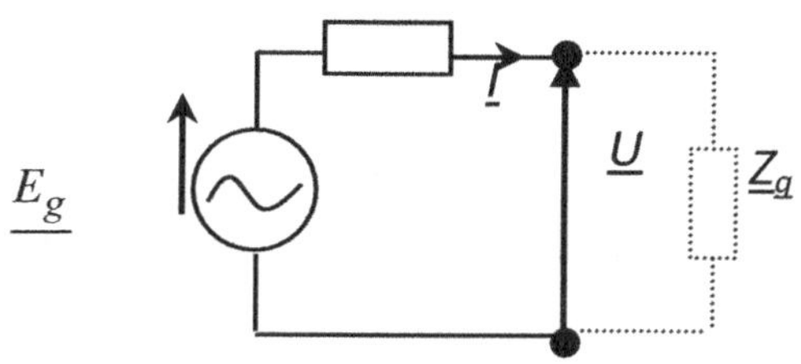

Figure 3.1. *Thévenin equivalent diagram of the generator*

The mean power dissipated into the load is:

$$P = \frac{\left(\underline{U}\,\underline{I}^* + \underline{U}^*\underline{I}\right)}{4}$$

By setting $\underline{Z} = \mathrm{R} + \mathrm{jX}$ and $\underline{Z}_g = \mathrm{R}_g + \mathrm{jX}_g$:

$$\underline{U} = \frac{\underline{Z}}{\underline{Z} + \underline{Z}_g}\underline{E}_g$$

and:

$$\underline{I} = \frac{\underline{E}_g}{\underline{Z} + \underline{Z}_g}$$

Hence:

$$P = \frac{1}{4} \frac{\underline{Z} + \underline{Z}^*}{\left|\underline{Z} + \underline{Z}_g\right|^2} \left|\underline{E}_g\right|^2$$

We find:

$$P = \frac{1}{2} \frac{R}{\left(R + R_g\right)^2 + (X + X_g)^2} \left|\underline{E}_g\right|^2 = \frac{R}{\left(R + R_g\right)^2 + (X + X_g)^2} E_{g_{eff}}^2$$

To maximize the power, we need to minimize the denominator by taking $X = -X_g$ and look for the maximum of the function:

$$f(R) = \frac{R}{\left(R + R_g\right)^2}$$

In order to do so, let us calculate the derivative of that function:

$$\frac{d}{dR}\left[\frac{R}{\left(R + R_g\right)^2}\right] = \frac{R_g - R}{\left(R + R_g\right)^3}$$

This derivative takes a value of 0 for $R = R_g$, which means that a generator with internal impedance $\underline{Z}_g$ and whose rms emf is E_{rms} will give maximum power to its load Z when the load's impedance is equal to the complex conjugate value of its internal impedance – i.e.: $\underline{Z} = \underline{Z}_g^*$, and the maximum power is:

$$P_{MAX} = \frac{E_{eff}^2}{4R}$$

3.2.2.1.1. Case where the internal impedance and the load are pure resistances

Power dissipated by the generator:

$$P_g = E_{g_{eff}} I_{eff} = \frac{E_{g_{eff}}^2}{R + R_g}$$

Power received by the load:

$$P_r = RI_{eff}^2 = \frac{RE_{g_{eff}}^2}{(R + R_g)^2}$$

Yield:

$$\eta = \frac{P_r}{P_g} = \frac{R}{R + R_g}$$

NOTE.– In conditions of power matching, the energy yield is only 50% (see Figure 3.2, showing the variations in power and yield). The energy source supplies a total amount of power, half of which is dissipated into the load; the other half is lost by Joule effect in the generator's internal resistance.

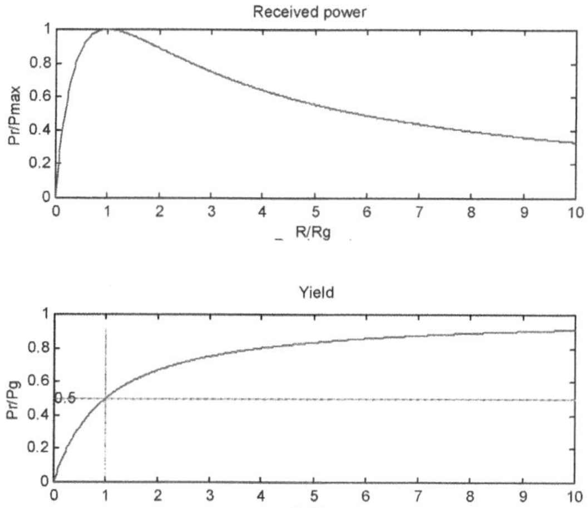

Figure 3.2. *Variations in power and yield*

For example, if the source is an integrated circuit of R_ic_out = R_gene = x ohms, then the load must be equal to x ohms for the best match.

Hence, it is crucial to know the value R_ic_out of the integrated circuit, and it is essential in order to find out how much of the available power can actually be transmitted to the load.

EXAMPLE.– For instance, the application note for integrated circuit NFC PN 532 from NXP presents the curve P = f(R) (see Figure 3.3) in its use in differential generator mode. The curve is symmetrical, and indicates two things:

1) In the pure case of production of the magnetic field H_max, the optimum value of R_match_differential is 35–40 Ω (at the peak of the curve).

2) As we can see from the curve, the maximum power zone is fairly flat, and, therefore, we do not lose a great deal of power as we approach higher values of "matching" resistances. Thus, with the same power supply voltage, the higher the sum of the internal resistance R_ic_out and the matched load, the lesser is the value of the current and the weaker is the power drawn from the supply. Hence, if we wish to optimize consumption from the direct componante (DC) supply of the integrated circuit, while retaining a worthwhile value of H_max, the value of R_match_ differential to take into consideration is closer to 60–70 Ω.

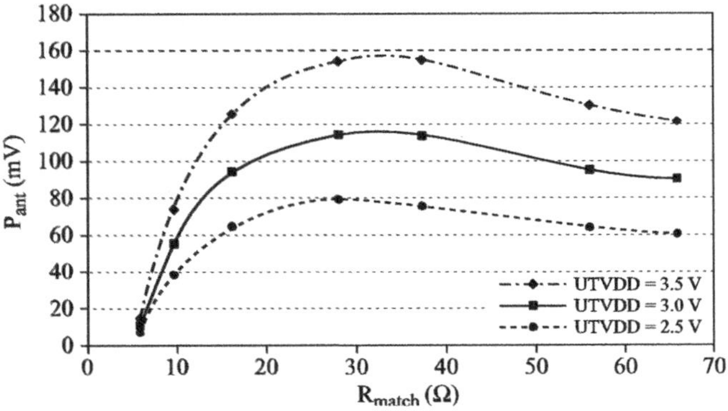

Figure 3.3. *Curve P = f(R)*

Thus, since the manufacturer cannot know, *a priori*, what the end user wants to do, they generally *recommend* an (average) value of matching impedance of around *R_match_ differential* = ~ *50Ω*.

3.2.2.2. *A little stronger*

For the form, let us linger a little longer in the pure sinusoidal regime, and add in a serial circuit with the two resistances of identical values R_gene_out = R_load of the previous network a serial circuit of type L, C, (where R_L = 0), and tune the latter circuit strictly to the incident frequency.

Figure 3.4 and those following it indicate, for example, the simulations using "ADS", from Agilent Technologies, of a 13.56 MHz system where, for example, $L = 1\mu H$ and $C = 137.6pF$, in which we apply a sine wave voltage with reference 1V_p, i.e. $2V\_pp$.

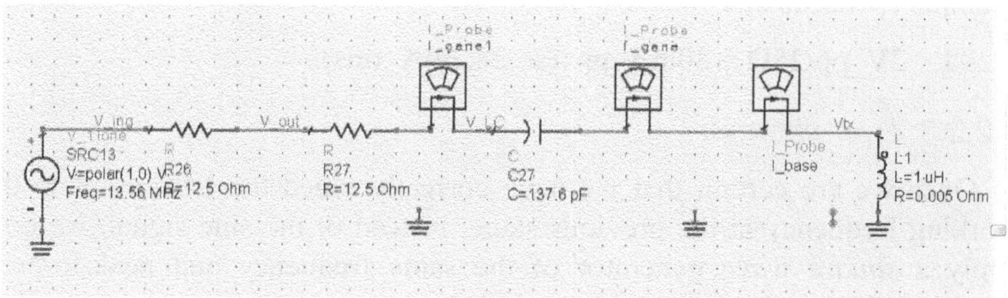

Figure 3.4. *Simulations by "ADS" from Agilent Technologies. For a color version of the figure, see www.iste.co.uk/paret/antenna.zip*

As indicated by Figures 3.5(a–c), *still using a sine wave*, only at that frequency, the voltage at the terminals of the serial LC circuit V_LC = 0 (Figure 3.5(c)), and the overall circuit is equivalent to the above two serial resistances, so in the regime established V_out = V_in/2 (Figure 3.5(b)).

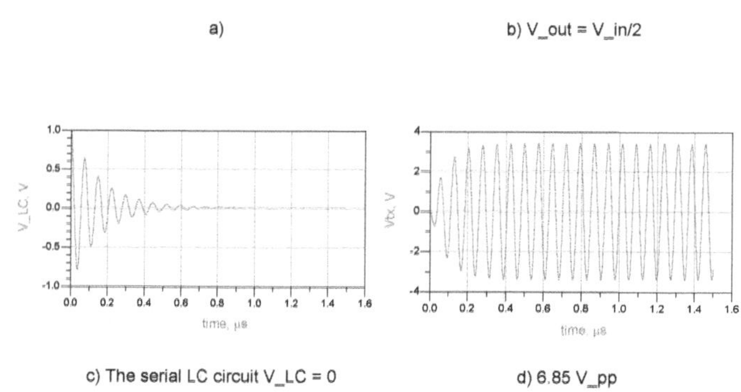

a) b) V_out = V_in/2

c) The serial LC circuit V_LC = 0 d) 6.85 V_pp

Figure 3.5. *Voltage at the terminals of the circuit LC*

EXAMPLE.– When R_gene = 12.5 ohms:

– Q = Lω/R = 1 × 85.72/(12.5 + 12.5) = 3.428;

– V_tx = Q V_in = V_in × L ω/(R+R) = 6.85V_pp (Figure 3.5(d)). In the example (…because V_in = 2V_pp);

– I = 2V_pp/25Ω = 80mA_pp (so 28.2mA_rms).

3.2.2.3. *Ever stronger*

Once we are certain that we have correctly tuned the LC circuit to the working frequency at the previous stage, instead of the sine signal, we now apply a *square wave* generator of the same frequency and peak-to-peak amplitude V_ing. In the new example below, we choose an inductance $L = 2\mu H$ (i.e. a capacity $C = 68.8pF$ to preserve the same tuning as before), a source which delivers a (new) no-load signal of V_ing = *3.3V_pp* (between 0 and 3.3 V). Why this latest value? Because it is this value of voltage which, in a *single-ended* regime, gives us a conventionally powered integrated circuit at V_batt = 3.3 V (see Figure 3.6).

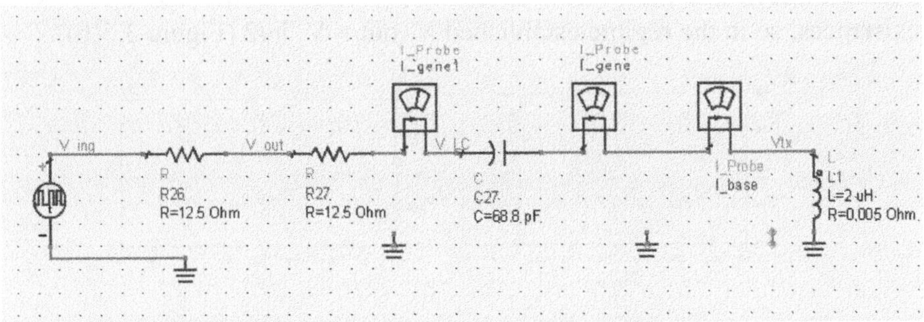

Figure 3.6. *Simulations by "ADS" from Agilent Technologies. For a color version of the figure, see www.iste.co.uk/paret/antenna.zip*

3.2.2.3.1. What is the value of the current in the circuit? Good question

If the serial circuit (L, C) were a short-circuit at all frequencies in the spectrum (…which is absolutely not the case), the current circulating in the resistors would be a perfectly square wave, and its peak-to-peak amplitude would be:

I_pp = V_batt/(R_int + R_load)

or:

I_p = V_batt/2 (R_int + R_load)

Unfortunately, the circuit L, C (a second-order filter with a slope of –12 dB/octave) has infinite frequency selectivity only at its tuned frequency of 13.56 MHz, and the total impedance (R + j X) of the network R, L, C is resistive (where R is real and X = 0) only at that frequency. The value of its impedance is significantly different at other frequencies, and the network L, C does not represent a true bandpass filter, so:

– given the spectral content of the square-wave signal, each of the harmonics will be affected differently;

– because of the selectivity of the circuit L, C tuned to 13.56 MHz, for the most part, it only allows passage to the component of the current representing the fundamental frequency of the square-wave signal and, by virtue of the superposition theorem, it is possible to separate the study of each harmonic in the Fourier decomposition of that signal – *in our case, we can consider solely that* the fundamental component of the wave *N = 1, for which we obtain a correction factor of 4/π = 1.274..., so greater than 1. The AC circulating in the output pin of the source will then be considered to have a (*nearly*) sinusoidal shape* (and will contain only very few of the first harmonics of the decomposition of the signal passing through the tuned circuit).

REMINDER.– Depending on the origin of the times chosen, the Fourier series decomposition of the generator's theoretical square-wave signal of the output stage (integrated circuit) is written thus:

$$E_{TX1} = \frac{E_0}{2} + E_0 \frac{4}{\pi}\left\{\cos(\omega t) + \frac{1}{3}\cos(3\omega t) + \frac{1}{5}\cos(\omega t) + \ldots\right\}_c$$

and the (peak) amplitude of the N-order spectral component has the value:

$$E_{TX\,N} = \frac{4}{\pi N} E_0$$

EXAMPLE.– In the present case, with a matched load of 12.5 ohms (thus maximum power transfer to the load, R_in = R_ load), the peak-to-peak output current I_out_pp of the integrated circuit will then be (supposing that the filter only allows passage to the sine wave of the fundamental frequency):

$$I\_out\_pp = (4/\pi) \times (V\_batt/(R\_int + R\_load\_match))$$

and the peak current I_out_p_max will thus be equal to half (see Figure 3.7):

$$\begin{aligned} I\_out\_p &= (4/\pi) \times (V\_batt/2/(R\_int + R\_load\_match)) \\ &= (4/\pi) \times (3.3/2/(12.5 + 12.5)) \\ &= 84.08mA\_p \end{aligned}$$

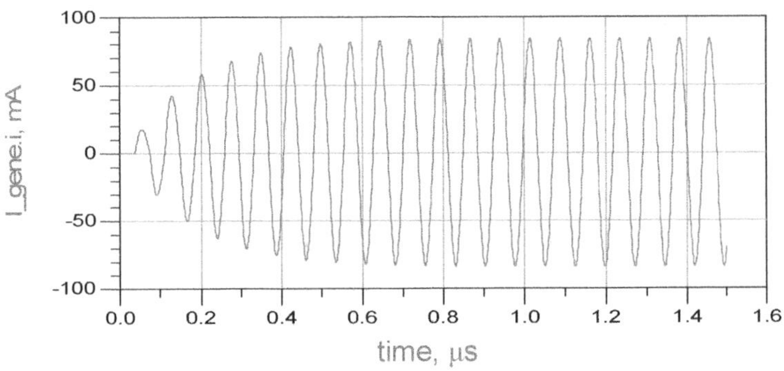

Figure 3.7. *Peak current I_out_p_max*

3.2.2.3.2. What about the output voltage from the integrated circuit?

The voltage V_out (i.e. the voltage which is actually available from the integrated circuit's output pins), contrary to expectations, is not a neat little square-wave signal with amplitude V_ing/2 (half the value of the no-load amplitude – i.e. 3.3V/2 = 1.65V_pp, because the load is matched), but instead the "unpredictable" signal shown below, because the whole network R, L, C actually forms a bandpass filter for all the spectral components of the incident square-wave signal, *but* whose average value of the peak-to-peak amplitude is (of course) our well-known V_ing/2 = 3.3/2 = 1.65 Vpp (see Figure 3.8).

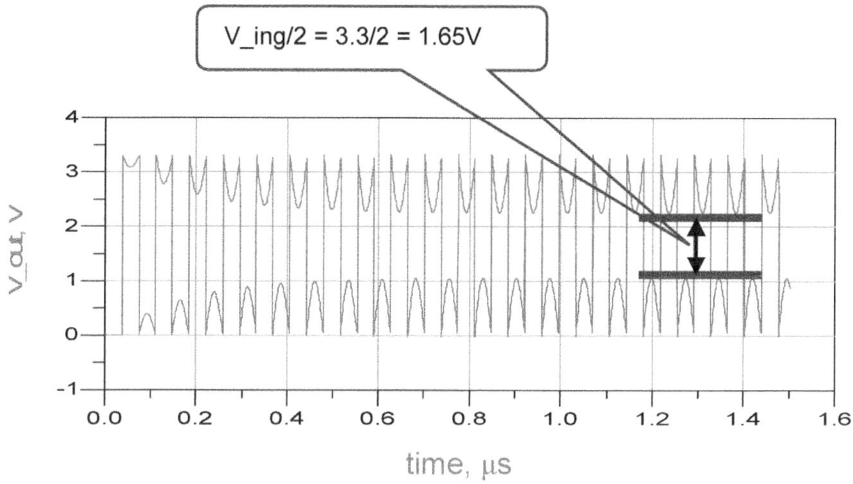

Figure 3.8. *Voltage V_out*

3.2.2.3.3. What about the voltage at the terminals of the L, C circuit?

Readers might have expected this voltage to be flat, and equal to zero, as in the previous case. However, it is not so. Although the average value is zero at around 1.65 V, the instantaneous value is not. The terrible pattern we are left with is the residual of all the high-order 1/N harmonic components of the square-wave signal, poorly filtered by the L, C circuits, but which is only properly tuned to the fundamental frequency of 13.56 MHz (see Figure 3.9).

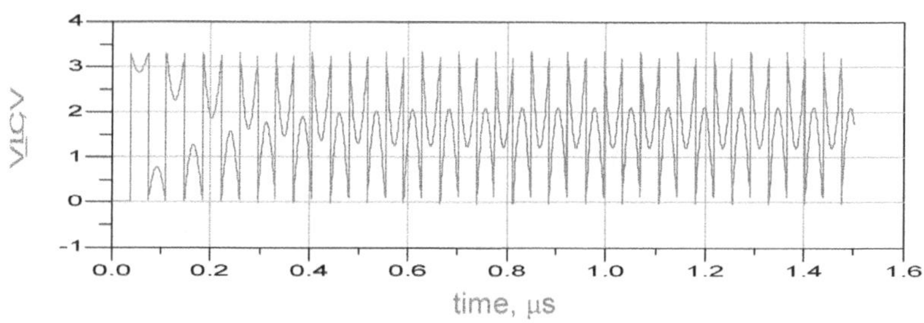

Figure 3.9. *Voltage at the terminals of the whole L, C circuit*

... and that at the terminals of the inductance L (Figure 3.10).

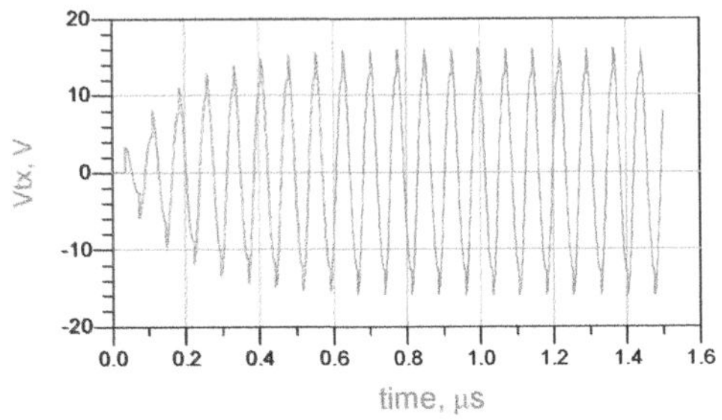

Figure 3.10. *Voltage at the terminals of the inductance L*

It is quite simple: VL_rms = (L ω × I_rms) – no more, no less.

In the wake of this necessary didactic interlude, let us now move on and look at the reality.

3.2.3. *Choice of integrated circuit*

Suppose that, for reasons of availability, cost, etc., we have chosen as our integrated circuit, for example, the circuit PN 532 for NXP. Its data sheet states:

– V battery (VBAT) battery supply voltage: 2.7–5 V;

– VBAT battery nominal supply voltage: 3.4 V (often this is powered at *V_bat = 3.3 V nominal)*;

– the DC consumed on VBAT = 3.4 V: ITVDD transmitter supply current continuous wave: typic 60 and max. 100 mA, with a note indicating: typical value using a complementary driver configuration and an antenna matched to 40 ohms between TX1 and TX2 at 13.56 MHz.

In addition, the Application Note AN 1445 for the same circuit recommends using "*Topology 1*" indicated below (known as a "symmetrical" attack, or indeed "differential mode", which is the most conventional). As we will see in the next section, this type of setup facilitates greater power transfer to the load.

Let us briefly recap the well-known differences between these application setups.

3.2.3.1. *"Single-ended" output*

In the case of *single-ended* output (the pins Ant 1 and Ant 2 are together and the mass, as indicated (Figure 3.11), the voltage V_out_pp as a square wave applied to the antenna network can be no greater than the circuit's supply voltage Vb and the output impedance/internal resistance of the integrated circuit has a certain value "R_int_single-ended".

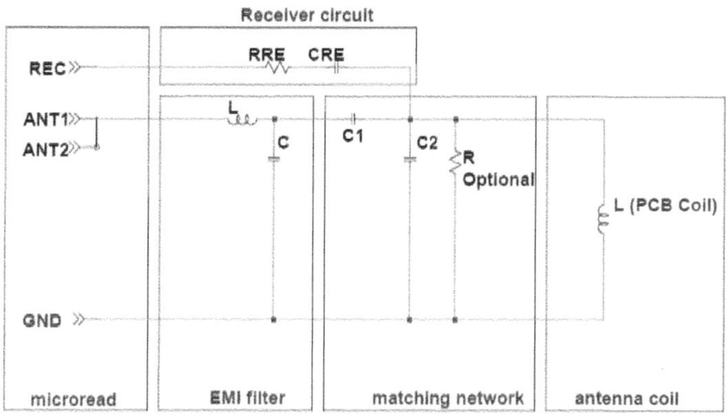

Figure 3.11. *"Single-ended" output*

The rms power available to a matched load equal to R_int_single-ended is then:

$$P\_single\text{-}ended = ((4/\pi \times Vb/2)/\sqrt{2}/2)^2/R\_int\_single\text{-}ended$$

$$= (Vb^2/2\pi^2)/R\_int\_single\text{-}ended.$$

Of course, the same amount of power is dissipated into the generator.

3.2.3.2. *"Differential mode" output*

In the case of a setup with a *symmetrical* output (differential mode – see Figure 3.12 (meaning between pins Ant 1 and Ant 2), the two internal output stages Ant 1 and Ant 2 then operate as a so-called *"H bridge"* (... the mass is then merely an arbitrary reference for the potential).

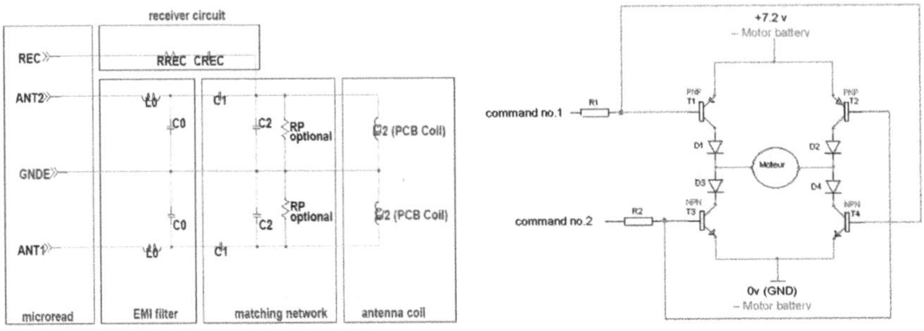

Figure 3.12. *"Differential mode" output*

This enables us to double the value of the peak-to-peak voltage applied to the setup – i.e. to obtain a voltage V_out_pp applied to the whole of the network equal to double the circuit's supply voltage – i.e. 2 Vb – of course on an internal impedance which is double 2R_int_single-ended in differential mode instead of R_int_single-ended in single-ended mode.

The rms power available in differential mode on a matched load equal to 2R_int_single-ended, therefore, is:

$$P\_differential = ((4/\pi \times Vb)/\sqrt{2}/2)^2/2R\_int\_single\text{-}ended$$

$$= (Vb^2/\pi^2)/R\_int\_single\text{-}ended$$

which is double the value of P_single-ended.

Of course, the same amount of power is dissipated into the generator.

NOTE.– Unless the cost of the filtering – and matching components – used then doubles, this setup exhibits two notable advantages:

– it doubles the available power, as explained above, thus offers a potential increase in the working distance;

– taking the *mass* as a reference potential, this setup renders symmetrical the radiation of the electrical fields E present at the extremities of the antenna coil, and thereby nullifies their direct effects on EMC pollution, all without altering the production of magnetic field H created by the coil and necessary for the application. See Figure 3.13.

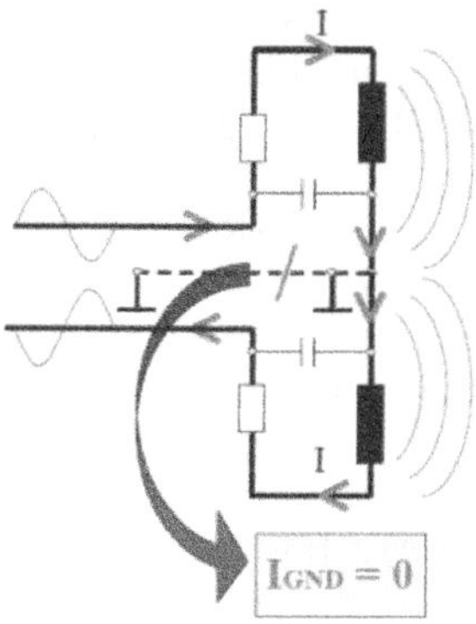

Figure 3.13. *The radiation of the electrical fields E becomes symmetrical*

3.2.3.3. *Conclusions*

Single-ended setup
internal resistance of single-ended generator = R_int_single-ended

$$I\_out\_p\_rms = [(4/\pi) \times (V\_batt/2/(R\_int\_single\text{-}ended + R\_load\_match))]/\sqrt{2}$$

or, indeed, with R_load_match = R_int_single-ended:

$$I\_single\text{-}ended\_rms = (4/\pi) \times [(V\_batt/2/(2\ R\_int\_single\text{-}ended))]/\sqrt{2}$$

$$I\_single\text{-}ended\_rms = (V\_batt/R\_int\_single\text{-}ended)/(\pi\ \sqrt{2})$$

Essentially, regardless of the application, as long as the load is matched:

Set R_load = R_in_single-ended and X = 0 at 13.56 MHz

The circuit's output will always be the same rms current.

And the power in the matched load R_load_match = R_int_single-ended will be equal to:

$$P\_rms\_single\text{-}ended = R\_int\ I\_\_single\text{-}ended\_rms^2 = R\_int\ [(V\_batt/R\_int\_single\text{-}ended)/\pi\ \sqrt{2}]^2$$

$$P\_rms\_single\text{-}ended = V\_batt^2/(2\ \pi^2\ R\_int\_single\text{-}ended)$$

Differential/symmetrical setup

internal resistance of a differential generator = R_int_differential = 2 R_int_single-ended

$I\_out\_p\_rms = [(4/\pi) \times (V\_batt/(R\_int\_differential + R\_load\_match\_differential))]/\sqrt{2}$

or, indeed, with R_load_match_differential = R_int_differential = 2 R_int_single-ended:

$I\_differential\_rms = (4/\pi) \times [(V\_batt/(2\ R\_int\_differential))]/\sqrt{2}$

$I\_differential\_rms = (V\_batt/R\_int\_single\text{-}ended)/(\pi \sqrt{2})$

$I\_differential\_rms = I\_single\text{-}ended\_rms$

Essentially, regardless of the application, as long as the load is matched:

Set R_load_differential = R_int_differential and X = 0 at 13.56 MHz

The circuit's output will always be the same rms current.

And the power in the matched load R_load_match_differential = R_int_differential will be equal to:

$P\_differential\_rms = R\_int\_differential\ I\_differential\_rms^2 = R\_int\_differential\ [(V\_batt/R\_int\_single\text{-}ended)/\pi \sqrt{2}]^2$

or R_int_differential = 2 R_int_single-ended

$P\_differential\_rms = 2\ R\_int\_single\text{-}ended\ [(V\_batt/R\_int\_single\text{-}ended)/\pi \sqrt{2}]^2$

$P\_differential\_rms = V\_batt^2/(\pi^2\ R\_int\_single\text{-}ended)$

$P\_differential\_rms = 2\ P\_asm\_rms$

Box 3.1. *Summary of "single-ended versus differential" setups*

Example with the circuit PN 532 from NXP:
V_batt = 3.3V
Upon matching

Single-ended	Differential
V_out_pp = 3.3V R_match_single-ended = 17.5Ω I_rms = 42.4mA_rms	V_out_pp = 6.6V R_match_differential = 35Ω I_rms = 42.4mA_rms
P = 31.55mW	P = 63.11mW

Box 3.2. *Examples of "single-ended versus differential" setups*

3.2.4. *Legislational constraining aspects and EMC pollution*

In radiofrequency applications (RFID, NFC and C°), one of the most important constraints is to respect the threshold levels set by national/international legislation on radiation and RF pollution (see Chapter 2). It must be remembered that for export to the United States, the regulations Federal Communications Commission (FCC) 47 part 15 are essentially the same as Europe's European Telecommunications Standards Institute (ETSI) 300 330.

Let us now look at this a little more closely.

At the frequency of 13.56 MHz, the regulations stipulate that the spectral pattern mask given in Annex L of ETSI 300 330 be respected (see Figure 3.14).

The value of the carrier at 13.56 MHz +/– 7 kHz is not too difficult to regulate: we simply place a quartz crystal in the application.

The most difficult thing is to limit (reduction by 60 dB) the levels of the odd-numbered harmonics of high rank (produced by the origin of the square-wave signals, produced by the integrated circuit in use, and often considered to be "spurious"), which fall in the RF bands (generally of order N = 13 = 176.28 MHz at N = 27 = 366.12 MHz).

For this reason, we need to use a filter.

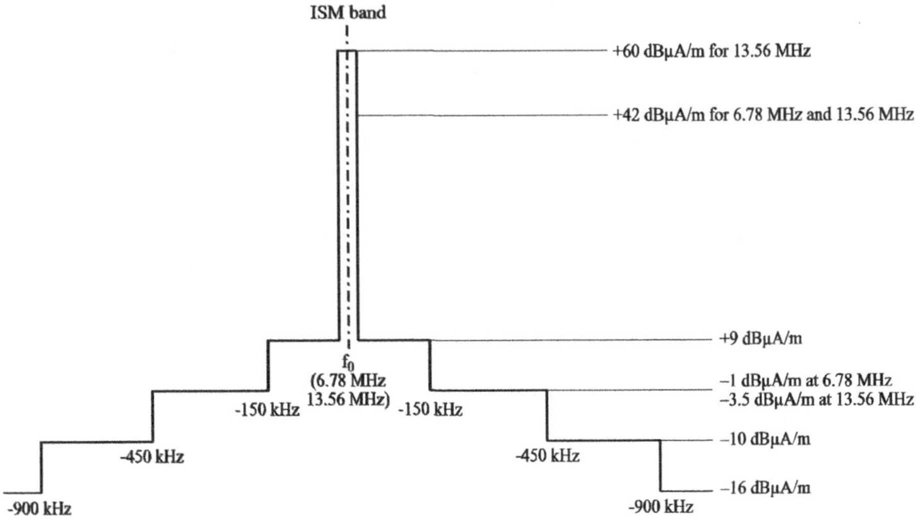

Figure 3.14. *Spectral pattern from Annex L of ETSI 300 330*

3.2.5. *EMC filtering*

3.2.5.1. *Filtering circuit*

In order to satisfy all the above regulations, before designing the impedance-matching circuits between a "signal generator" and the "final antenna circuit", it is necessary to interpose a low-pass filter (for reasons of simplicity and of cost, type L C, as indicated in Figure 3.15) in order to filter out the harmonics contained in the output signal which, remember, are only the odd-numbered harmonics 3, 5, 7, 9, etc., at 13.56 MHz.

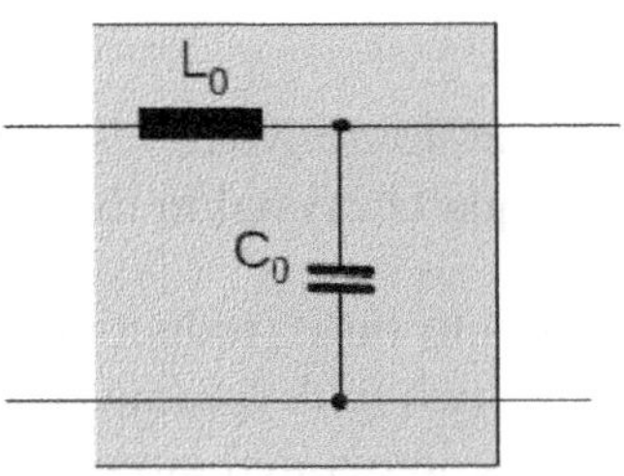

Figure 3.15. *Low-pass filter*

This filtering circuit primarily serves two functions:

1) filtering of the signal, as indicated above;

2) transformation of the impedance, as indicated in a number of section, with the secondary goal being to reduce the rise time after a phase modulation (in reader mode) and to increase the reception bandwidth.

IMPORTANT NOTES.–

– The main problem with this filter is that it must be physically "realizable and reproducible". This means that the values of Lo and Co must not be ludicrously low, so as to prevent parasitic effects (capacities) – e.g. Lo = 270nH-2µH.

– Generally, in concrete terms, in order to perform this filtering, we use a setup with a differential topology, as shown in Figure 3.16, by physically dividing the inductance Lo above into two inductances with identical values L to that given above (thus, we have a total value of Lo = 2L, because it is a serial circuit), and splitting the capacity C into two capacities arranged as a series of values C (thus, the serial circuit gives us a total of = Co/2) … which changes nothing in terms of the match calculated earlier, because 2L × C/2 = Lo Co.

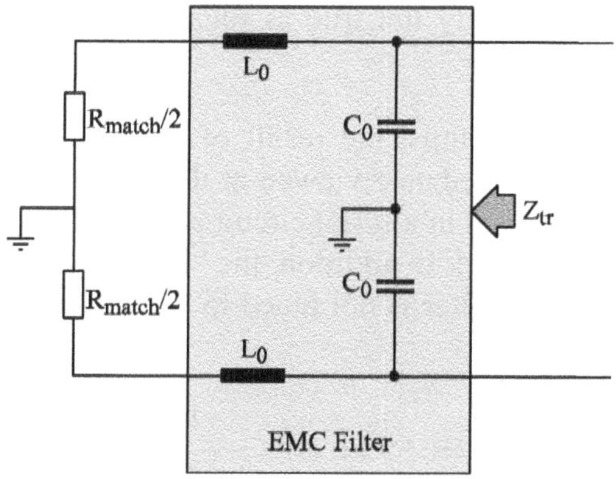

Figure 3.16. *Filtering with a differential setup*

It should be noted that the inductance should be capable of withstanding the passage of the values of rms currents estimated above (e.g. of the order of magnitude of ~60–80 mA_rms max).

3.2.5.2. *Tuning frequency of the filter*

Evidently, the bandwidth of this filter needs to be such that it can correctly allow through the highest frequencies of the signal which the initiator wishes to transmit, and cut out the harmonics to the fullest extent possible.

For this reason, we generally choose to tune that low-pass filter at least to the value of the fundamental frequency of the carrier (square-wave signal at 13.56 MHz), increased by the highest value of the fundamental frequency of the subcarrier – the *upper side band* (square-wave signal with maximum throughput of 848 kbits/s superposed on the carrier), which gives a minimum total of 14,608 MHz. This value is generally rounded up to a slightly higher value of 15.5–17 MHz.

EXAMPLE.– The cutoff frequency of a second-order L, C filter is situated at $f = 1/2\pi \sqrt{(LC)}$. For its part:

NXP recommends	f = 13.65 MHz;	Lo = 1 μH;	Co = 136 pF
	f = 15.5 MHz;	Lo = 560 nH;	Co = 180 pF
Inside Secure, etc.	f = 14.4 MHz;	Lo = 220 nH;	Co = 560 pF

...but the same is absolutely not true as regards impedance matching (see below).

Hence, in the above setups, the result of the filter's output impedance equation Z_out_filter (impedance viewed at the output of the filter; the two elements (R + Lo) arranged in a serial circuit are parallel to Co) is a complex variable of the type (a + j b). In addition, the "b" part of (a + jb) cannot have a value of 0, because the filter is not tuned to the carrier. Let us express that in equation form:

$$Z\_out\_filter = \frac{(R + j\,Lo\,\omega) \times \dfrac{1}{j\,Co\,\omega}}{(R + j\,Lo\,\omega) + \dfrac{1}{j\,Co\,\omega}} = (a + j\,b)$$

$$Z\_out\_filter = \frac{(R + j\,Lo\,\omega)}{(1 - Lo\,Co\,\omega^2) + (j\,R\,Co\,\omega)} = (a + j\,b)$$

We can multiply the top and bottom of this equation by the conjugate of the denominator. After reduction of the terms, we obtain the following literal final equation:

$$Z\_out\_filter = \frac{R}{(1 - Lo\,Co\,\omega^2)^2 + (R\,Co\,\omega)^2} + j\,\omega\,\frac{Lo\,(1 - Lo\,Co\,\omega^2) - R^2\,Co}{(1 - Lo\,Co\,\omega^2)^2 + (R\,Co\,\omega)^2} = (a + j\,b)$$

Let us identify these terms, one at a time (by setting R = $\frac{R_{match}}{2}$ for a differential setup). Thus, we obtain:

$$R_{tr} = \frac{R_{match}}{\left(1-\omega^2\cdot L_0\cdot C_0\right)^2+\left(\omega\cdot\frac{R_{match}}{2}\cdot C_0\right)^2}$$

$$X_{tr} = 2\cdot\omega\cdot\frac{L_0\cdot\left(1-\omega^2\cdot L_0\cdot C_0\right)-\frac{R_{match}^{\ 2}}{4}\cdot C_0}{\left(1-\omega^2\cdot L_0\cdot C_0\right)^2+\left(\omega\cdot\frac{R_{match}}{2}\cdot C_0\right)^2}$$

Of course, we then wish to transfer the maximum possible amount of power to the final load (the antenna). In order to do so, we need to couple that impedance Z = R + jX with its conjugate impedance Z* = R – jX so as to cancel out the imaginary term (see Figure 3.17). This will be the main aim of the matching circuit, which we will examine later on.

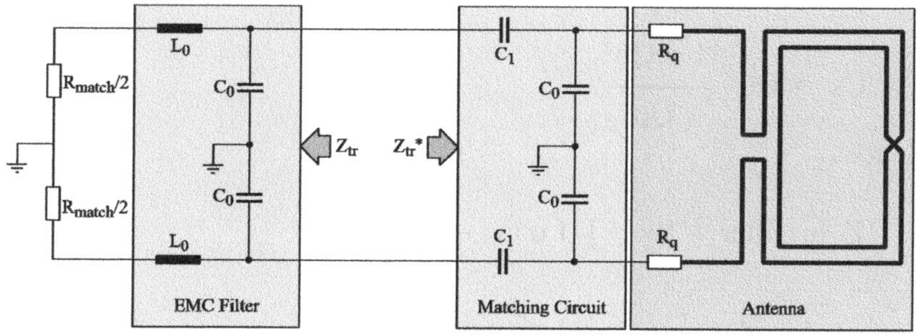

$Z_{tr} = R_{tr} + jX_{tr}$
$Z_{tr}^* = R_{tr} - jX_{tr}$

Figure 3.17. *Matching circuit*

3.2.5.3. *Condition for transferring maximum watt power to a load Rp*

Here, we come to another important aspect of the same problem. Let us now define the resistance *Rp* (see Figure 3.18) as being the pure equivalent resistive load that the filtering circuit Lo, Co can withstand, when it is parallel to the capacity Co.

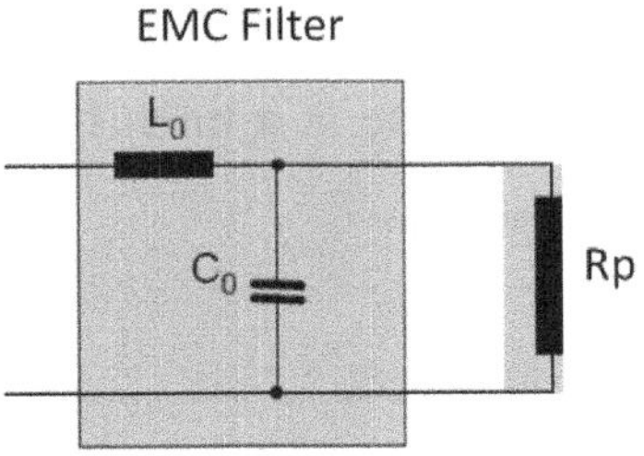

Figure 3.18. *Pure equivalent resistive load*

3.2.5.3.1. Calculation of the impedance "Z_in_filter"

Let us calculate the input impedance Z_in_filter of the whole circuit formed by Lo in a series with Co, and its parallel load Rp. To do so, using the conventional formulae for transforming a serial circuit into a parallel one, and vice versa, we will transform the parallel network (Co, Rp) into a serial network (Rs, Cs). We have the following new parameter:

$$Qp = Rp\ Co\ \omega$$

with:

$$Rs = \frac{Rp}{1 + Qp^2}$$

and:

$$Cs = Co\ \frac{1 + Qp^2}{Qp^2}$$

Hence:

$$Z\_in\_filter = Rs + j\left(Lo\ \omega - \frac{1}{Cs\ \omega}\right)$$

In regard to the output resistance R_ic_out of the integrated circuit (the generator), at 13.56 MHz, in order to transfer the maximum power into that new resistance Rp, the impedance Z_in_filter = a + jb of the whole circuit Lo series to Co and its parallel load Rp must behave as follows (as usual):

– firstly, in order for the load Rp to receive the maximum watt power (the other components being reactive), the value of the input impedance Z_in_filter must only be real, and also because that it behaves like a pure resistance, its complex part "b" must have the value of 0. *This means that the filter Lo, Cs must necessarily be tuned to the value of the operating frequency* (meaning that this filtering is applicable to the system, but very often the case), so:

$$\text{Lo Cs } \omega^2 = 1$$

Thus:

$$\text{Cs} = 1/\text{Lo } \omega^2$$

and therefore, in this case:

$$\text{Z\_in\_filter} = \text{Rs}$$

– furthermore, its value "a" = Rs;

– and finally, the condition of matching of the generator to its load must be fulfilled – in other words, the value of the equivalent serial resistance Rs of the filter must be equal to the generator's output internal resistance R_ic_out:

$$\text{Rs} = \text{R\_ic\_out} = \frac{\text{Rp}}{1 + \text{Rp}^2 \text{ Co}^2 \ \omega^2} = \frac{\text{Rp}}{1 + \text{Qp}^2}$$

or:

$$\text{Rp} = (1 + \text{Qp}^2) \text{ R\_ic\_out}$$

Having defined:

$$\text{Qp} = \text{Rp Co } \omega$$

$$= [\text{Rs } (1 + \text{Qp}^2)] \ [\text{Cs Qp}^2/(1 + \text{Qp}^2)] \ \omega$$

Hence:

$$1 = \text{Rs Cs Qp } \omega$$

$$Qp = 1/(Rs\ Cs\ \omega)$$

By replacing Cs with its value Cs = 1/Lo ω^2 (as we have deliberately tuned it to the working frequency), we find:

$$Qp = Lo\ \omega/Rs$$

Thus:

$$Rp = [\ 1 + Qp^2\ ]\ R\_ic\_out = [\ 1 + \left(\frac{Lo\ \omega}{R\_ic\_out}\right)^2\ ]\ R\_ic\_out$$

EXAMPLE 3.1.– With the integrated circuit PN 532 NXP and a differential setup:

f	= 13.56 MHz		
R_ic_out	= 35 Ω		
Lo	= 2 µH →	Cs = 1/Lo ω^2	= 69 pF
		Rp = above	= 863 Ω
		Qp = √(Rp/R_ic_out - 1) = 4.86	
		Co = Cs Qp²/(1 + Qp²) = 66.16 pF	

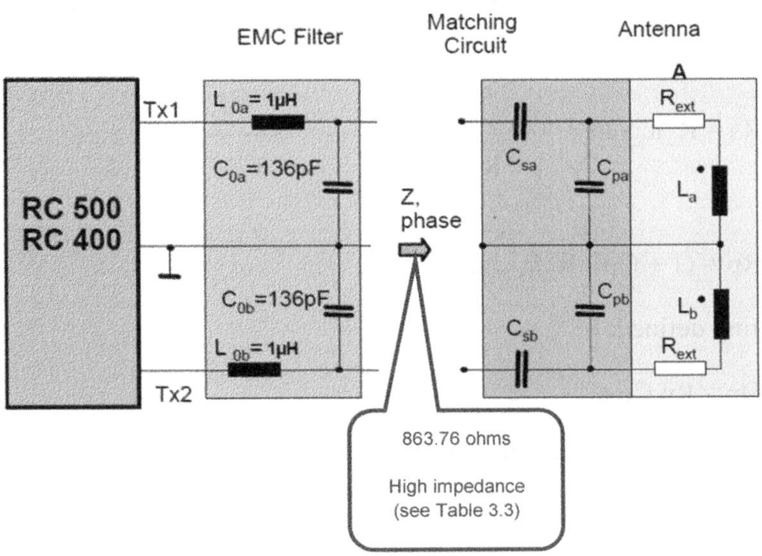

Figure 3.19. *Examples*

EXAMPLE 3.2.– (Table 3.3).

Examples			Inside Single-ended setup	NXP Differential setup See Figure 3.17	
Hypotheses	*Formulae*				
Working_freq		MHz	13.56	13.56	
ω	= 2 π f	rd/s	85.16	85.16	
R_ic_out		Ω	12.5	35	
Rs	= R_ic_out	Ω	12.5	35	
Chosen Lo value		μH	0.22	2	2 × 1μH in series
Freq_cutoff		MHz	13.56	13.56	
Cs for 13.56		pF	626.81	68.95	
Solutions					
Qp	= 1/ (Rs Cs ω)		1.50	4.87	
Qp^2			2.25	23.68	
$1 + Qp^2$			3.25	24.68	
Co	= Cs $Qp^2/(1+ Qp^2)$	pF	433.73	66.16	2 × 130 pF in series
Rp	= Rs $(1 + Qp^2)$	Ω	40.58 Low	863.76 High	

Table 3.3. *Examples*

3.2.5.3.2. Example of simulation of the filtering LC circuit on its own (R_ic_out = 12.5Ω – single-ended)

With the values taken from Table 3.3 (single-ended 12.5 Ω), Figures 3.20–3.22 show the results of simulations.

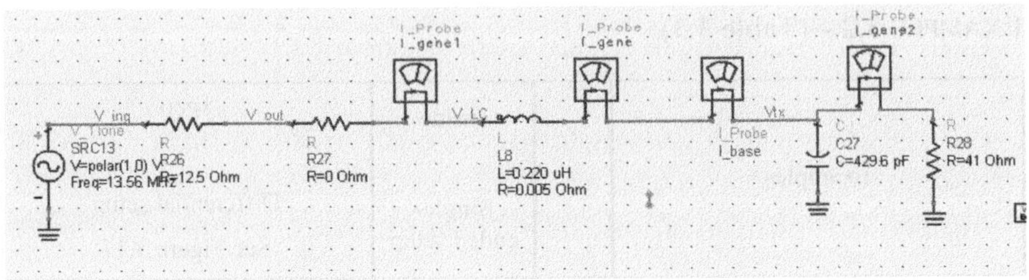

Figure 3.20. *Diagram of simulations. For a color version of the figure, see www.iste.co.uk/paret/antenna.zip*

The circuit is well-matched R_int = (R_apparent = R_in), so V_out = V_in/2.

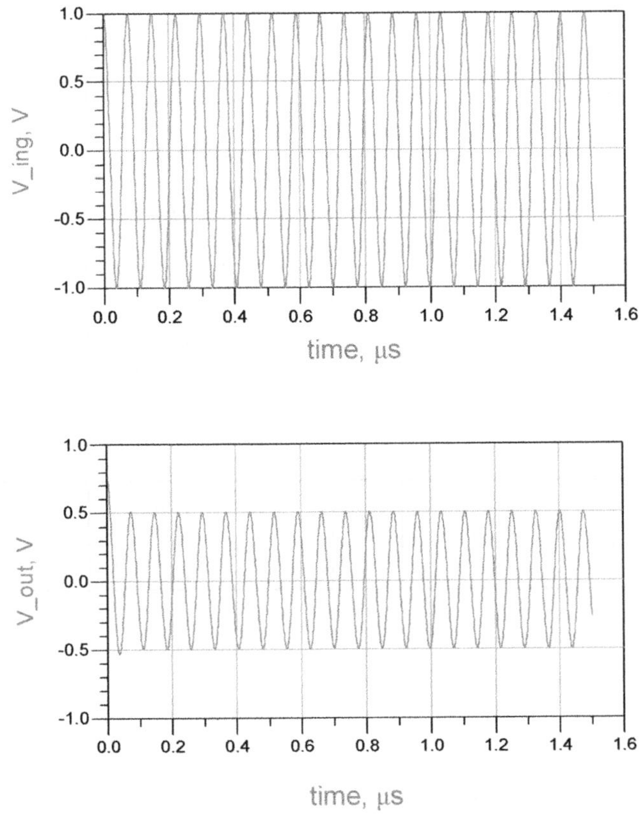

Figure 3.21. *Simulation results*

The current I_gene1 = V_gene/(R_int + (R_apparent = R_in)) = 1Vpp/(12.5 + 12.5) = 40 mA.

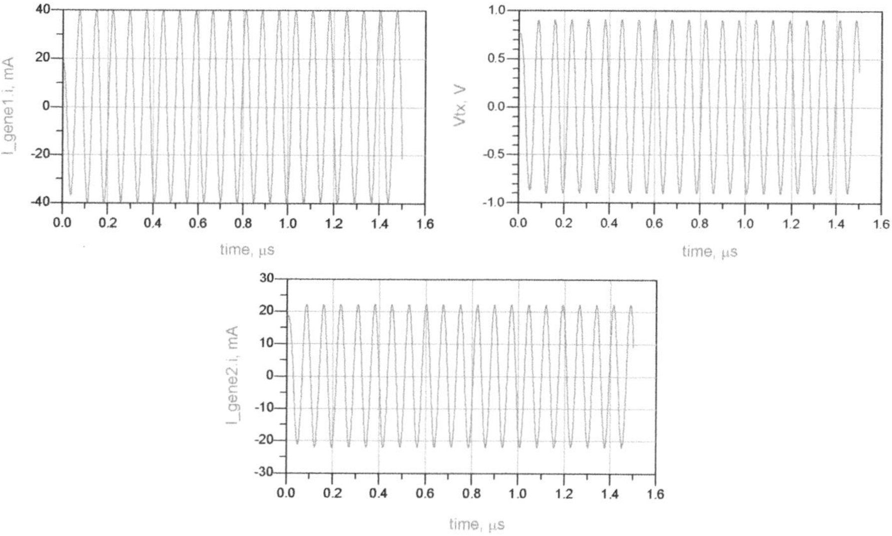

Figure 3.22. *Simulation results (cont.)*

3.2.5.4. *Current circulating in the resistance Rp*

It is important to know the value of the current circulating in the load Rp for the next stage of the process. In addition, it should be noted that the set formed by the capacity Co and the resistance Rp, on the one hand, and the inductance L, on the other hand, forms a dividing bridge whose potentiometric voltage ratio is:

$$\frac{V\_(Rp//Co)}{V\_network} = \frac{[Z(Co)//Z(Rp)]}{Z(Lo) + [Z(Co)//Z(Rp)]}$$

$$= \frac{Rp}{(Rp - Lo\ Rp\ Co\ \omega^2) + j\ Lo\ \omega}$$

However, we showed earlier that in order for the value of Z_in_filter = R_ic_out, then we must have Lo Cs $\omega^2 = 1$, and given that:

$$Cs = Co \frac{1 + Qp^2}{Qp^2} \quad \text{then} \quad Co = Cs \frac{Qp^2}{1 + Qp^2}$$

We find:

$$= \frac{Rp}{\dfrac{Rp}{1 + Qp^2} + j\, Lo\, \omega}$$

In addition: $Qp = Rp\, Co\, \omega$

$$Lo\, Cs\, \omega^2 = 1 \quad \rightarrow \quad Lo\, \omega = 1/Cs\, \omega$$

$$Cs = Co \frac{1 + Qp^2}{Qp^2}$$

$$1/Cs\, \omega = \frac{Qp^2}{Co\, \omega\, (1 + Qp^2)} = Lo\, \omega$$

Hence, we can write:

$$= \frac{Rp}{\dfrac{Rp}{1 + Qp^2} + j \dfrac{Qp^2}{Co\, \omega\, (1 + Qp^2)}}$$

$$= \frac{Qp\, (1 + Qp^2)}{Qp + j\, Qp^2}$$

$$= 1 - j\, Qp$$

Thus, finally, in terms of modulus:

$$= \sqrt{(Qp^2 + 1)} = \sqrt{(Qs^2 + 1)}$$

By setting "V_gene" as the rms equivalent sine-wave no-load voltage that should be furnished by the fundamental frequency at 13.56 MHz of the square-wave signal generator, i.e.:

$$V\_gene\_rms = (V\_batt/2 \times 4/\pi)/\sqrt{2}$$

In addition, having matched the impedances, the voltage present at the input to that network is equal to half the no-load voltage of the generator (V_network = ½V_gene), so:

$$V\_Rp\_rms = (\ \tfrac{1}{2}V\_gene\ ) \times\ (\sqrt{(Qs^2 + 1)})$$

and hence, now that we know V_Rp, we can easily deduce the current I_Rp in Rp:

$$I\_Rp\_rms = V\_Rp\_rms/Rp$$

$$I\_Rp\_rms = (\ \tfrac{1}{2}V\_gene\ ) \times\ (\sqrt{(Qs^2 + 1)})/Rp$$

3.2.5.5. *Power dissipated into the resistance Rp*

In matching, the power transmitted into the load Rp is, therefore:

$$P\_Rp\_rms = V\_Rp\_rms^2/Rp$$
$$P\_Rp\_rms = [\tfrac{1}{4}V\_gene^2 \times (Qs^2 + 1)]/Rp$$

Thus:

$$P\_Rp\_rms = [(\tfrac{1}{4}\ [(V\_batt/2 \times 4/\pi)/\sqrt{2}]^2 \times (Qs^2 + 1)]/Rp$$
$$P\_Rp\_rms = ((V\_batt^2/2\pi^2) \times (Qs^2 + 1))/Rp$$

which, in addition, upon matching, the load source impedance must be strictly equal to the power dissipated into the resistance R_int = R_ic_out, which for its part has the value:

$$\begin{aligned} P\_R\_int\_rms &= R\_int \times I\_rms^2 \\ &= R\_int \times ((V\_batt/R\_int)/\pi\ \sqrt{2})^2 \\ &= V\_batt^2/(2\pi^2\ R\_int) \end{aligned}$$

Put differently:

$$\begin{aligned} P\_R\_int\_rms &= V\_batt^2/(2\pi^2\ R\_int)\ = P\_Rp\_rms \\ &= ((V\_batt^2/2\pi^2)] \times (Qs^2 + 1))/Rp \end{aligned}$$

Thus, $V\_batt^2/(2\pi^2\ R\_int) = ((V\_batt^2/2\pi^2)] \times (Qs^2 + 1))/Rp$

$1/(R\_int) \qquad = (Qs^2 + 1)/Rp$

This means that, in matching, the relation between the resistances Rp and R_int must be equal to:

$$Rp = R\_int\ (Qs^2 + 1)$$

3.2.5.5.1. Overall summary for the "EMC filter" circuit

Examples			Inside	NXP	
			single-ended	Differential	
Hypotheses					
Ω		rd/s			
V_batt		V_dc	3.3	6.6	Bridge setup
R_ic_out		Ω	12.5	35	
Rs = R_ic_out		Ω	12.5	35	
Chosen Lo value		µH	0.220	2	
Cs for 13.56		pF	618.1	68.95	
Solutions					
Qp	1/ (Rs Cs ω)		1.5	2.87	
Qp^2			2.28	23.68	
$1 + Qp^2$			3.28	24.68	
Rp	$Rs\ (1 + Qp^2)$	Ω	41	863.76	
Co	$Cs\ (Qp^2/1+ Q^2)$	pF	429.6	66.16	
	See the simulated curves plotted above				
V_gene_no_load	$(V\_batt/2 \times 4/\pi)/\sqrt{2}$	V_rms	1.486	2.973	Rms sine transpose of square
V_network	1/2V_gene	V_rms	0.743	1.486	
$\sqrt{(Qp^2+1)}$			1.81	4.967	
V_Rp_rms	$½V\_gene\sqrt{(Qp^2+1)}$	V_rms	1.345	7.38	
I_Rp_rms	V_Rp_rms / Rp	mA_rms	32.8	8.545	
P_Rp_rms	$V\,I = Rp \times I\_Rp\_rms^2$	mW_rms	44	63	Max power is transferred into Rp

Table 3.4. *Overall summary*

IMPORTANT NOTE.– To conclude, all the above sections concerning EMC filtering are obviously valid only when the initiator integrated circuit delivers square-wave signals as output. It is quite clear that if it had output purely sinusoidal signals at 13.56 MHz, all of this (calculations, components L, C) could have been avoided. With a view to reducing costs and bulk, we can also put, before the output of the integrated circuit, a circuit to convert the square-wave signals into a pure sine-wave signal at 13.56 MHz by sampling the square wave, following by an analog converter. As we have shown, in this case, in order to have a hope of obtaining similar power in the load, it is necessary to have a supply voltage greater by the factor $4/\pi$ so as to compensate the lack of effect of the first harmonic of the Fourier decomposition.

3.2.6. *Choice of target used and incidence of its H_threshold*

3.2.6.1. *Estimation of the field H which needs to be produced by the initiator antenna*

In Chapter 1, we showed that the rationalized theoretical equation of the magnetic field H(a, r) produced by an initiator is of the form (Biot–Savart law):

$$H(a, r) = \frac{1}{[(1 + a^2)^{3/2}]} \times H(0, r)$$

with:

– "d" is the distance between the centers of the initiator/tag antennas;

– "r" is the radius of the initiator antenna;

– and $d = (a \times r)$.

In light of this equation, let us look at how to satisfy the application for a given target.

3.2.6.2. *Example of application for a specific target*

Suppose that the threshold field H_thres_worst_case of the target is equal to H_thres = 1A/m, and that the initiator antenna is a 19 × 19 cm square, so its equivalent radius is r = 10.72 cm.

– for a desired working distance of d = 30 mm;

– …so the value of "a" = d/r = 30/107.2 a = 0.28.

Using the table, we deduce that:

d	a	H_d/H_0
	0.28	~ 0.8

This indicates that to write this target at 3 cm, at the center of the initiator's coil, we need to have a magnetic field H_0 = approx. 1.25 A/m … which may be consistent with ISO 14443 and ISO 18092.

GENERIC EXAMPLE.– Table 3.5 is extremely important, and every designer should draw up a similar table before embarking on his/her own specific design. It summarizes and clearly indicates, for a given/desired format or antenna dimensions, what the minimum field strength H_0_min is that needs to be provided by an initiator in order to reach a desired distance "d" for a tag/target whose threshold field H_threshold is known or given or imposed. In the above example, the minimum guaranteed value for an NFC device is 1.5 A/m, in accordance with NFC standard ISO 18092, which is identical to the standard governing contactless chip cards: ISO 14443-2).

NOTE.– Additionally, the two right-hand columns show the same calculations for PCD, given as a reference in the conformity/test standard ISO 10373-6.

		Units	Symbols	Values				
				Ex. 1	Ex. 2	Ex. 3	PCD ISO	PCD ISO
Antenna	Circular radius	mm	circular				75	75
	External format	Mm	rectangular	50 × 10				
	Internal format	mm		45 × 8				
	Surface	mm²	s	500			17662.5	17662.5
	Equivalent radius	mm	r	12.6	12.6	12.6	75	75
Operating distance	Desired	mm	d	12.6	37.3	50	50	100
Field H	H_thres_worst case	A/m	H_thres = H(d)	1.5	1.5	1.5	1.5	1.5
	d/r		= a	1	3	3.96	0.666	1.33
	H_d / H_0			0.354	0.0316	0.0143	0.705	0.318
	H_0_min. required	A/m	H_0_min	4.24	47.47	104.90	2.21	4.72

Table 3.5. *Minimum field H_0_min of an initiator to achieve a distance "d"*

3.2.7. *Determining the inductance value of the initiator antenna*

3.2.7.1. *Value of maximum current in the initiator antenna*

In order to satisfy the magnetic field/magnetic induction (A/m or µT above) that the target needs in order to function properly at the distance required by the application, it is necessary to produce a magnetic field H whose value is directly linked to the values of the trio "N" (number of spires of the antenna), "I_ant" (current circulating in the antenna coil) and "r" (its equivalent radius). Remember, the field value at the center of the antenna is:

$$H\_0 = (N \times I\_ant)/2r$$

The maximum value of the current "I_ant" circulating in the initiator antenna will depend only on the application and on the possible power performances possessed by the leading (integrated) circuit in the antenna setup.

In summary, to begin with, for a given/imposed field H, we are free to choose the values of the trio "N", "r" (so the value of L) and "I_ant" depending on the available components and solutions, listed and discussed above. In addition, these three parameters are linked to one another by, first, the value of the inductance of the antenna (L = f(N and r)) and, second, the value of the antenna's quality coefficient Qs by way of its ohmic resistance Rs, which is equal to Qs = (L ω)/Rs, whose application requires a definite value (see the ISO standard) in order to work properly (timing, communication bandwidth, etc.). Furthermore, Rs itself depends on N and r. In short, everything reacts to everything else.

In addition, at the resonance, the antenna circuit Rs, L, C behaves like a pure resistance, and therefore consumes active power P (watts) in Rs, which is provided by the initiator's power platform (through all the matching networks described above which, as they are only reactive, do not consume power).

$$P = Rs \times I\_ant^2$$

By substituting the value of Rs from the first equation into the second, we obtain the important relation which links the current in the antenna and the other main parameters:

$$I\_ant = \sqrt{[(P \times Qs)/(L \times \omega)]}$$

We can often use this equation to look at the problem the other way around. Indeed, if we know the maximum power we have (or will have) – i.e. the maximum power which the initiator's integrated circuit is capable of delivering when matched), we can estimate the maximum value of the current for the initiator antenna I_ant, and then create an antenna for the initiator such that its inductance is compatible with the desired number of turns N with a radius "r" required to produce the magnetic field H which satisfies the application.

3.2.7.2. *In conclusion*

We know that when matched, the integrated circuit will deliver a maximum power P1 to its load of:

$$P1 = V\_out\_pp^2/(2\,\pi^2\,R\_int)$$

The current I_ant flowing through the antenna with the inductance L and total serial resistance Rs for a given Qs will, therefore, be:

$$I\_ant = \sqrt{(P/Rs)}$$

Hence, if we impose the match P1 = P, we can finally establish the relation linking the current in the antenna, as a function of the internal resistance of the integrated circuit used, and V_out _pp of the setup employed:

$$I\_ant = \sqrt{[\,((V\_out\_pp^2/(2\,\pi^2\,R\_int))/R1s\,]}$$

$$I\_ant\_rms = (V\_out\_pp/\pi\sqrt{2}) \times 1/\sqrt{(\,R1s\,R\_int)})$$

Thus, we have almost come to the heart of the problem. Beware, though: there is maximum current and maximum current. The purpose of the next section is to illuminate some of the subtle, hidden points concerning that maximum current. Let us focus, then, on the matter at hand.

V_batt = 3.3 V and f = 13.56 MHz, ω = 85.72 10^6 rd/s

Q1 = 35 (*mandatory* for ISO 14443 and ISO 18092-type applications)

On matching:

– single-ended:	*– differential mode:*
V_out_pp = 3.3 V	*V_out_pp = 6.6 V*
R_match_single-ended = 17.5Ω	*R_match_differential = 35Ω*
P_single-ended = 31.55mW	*P_differential = 63.11mW*
	P_differential = 2 P_single-ended

with L = 2.1µH (so R1s = 5.11ohms for Q = 35), we obtain:

I_ant = 79 mArms I_ant = 82.7 × √2 mArms I_ant = 111.2 mArms

(This is a shame. There were 20" options to obtain the correct value of the current, but not the filter elements to perform the right matching).

Box 3.3. *Examples with the circuit NXP PN 532*

At the center of the initiator's coil, the (final desired) value of the intensity of the magnetic field will be given by the relation:

$$H\_0 = (N \times I\_ant)/2r$$

where $I\_ant = \sqrt{[(P \times Qs)/(L \times \omega)]}$

We obtain $H\_0 = (N \times \sqrt{[(P \times Qs)/(L \times \omega)]})/2r$

By squaring the equation, we find:

$$H\_0^2 = (N^2 \times ([(P \times Qs)/(L \times \omega)]))/(2\,r)^2$$

or indeed:

$$L = (N^2 \times P \times Qs)/(H\_0^2 \times (2\,r)^2 \times \omega)$$

where, for instance:

- H_0 desired
= 1.25 A/m

- P1
= 44 mW

- quality coefficient Q1 "imposed" by the standards

ISO 14443 = 35

ISO 15693 = 35-100

- ω at 13.56 MHz = 2 × 3.14 × 13.56 × 10^+6

= 85.157 10^+6

Thus, for example, in the case of the initiator whose r is 10.2 cm:

$L1 = (N1^2 \times 0.044 \times 35)/(1.25^2 \times (2 \times 10.2\ 10^{-2})^2 \times (2 \times 3.14 \times 13.56 \times 10^{+6}))$

Thus, $L1 = N1^2 \times 0.278$ in µH

In short, this means that in order to satisfy our problem, is there a mechanical technology to make an antenna such that the couple of L and N values obtained satisfies this equation?

This is a good question, and one which is well-worth asking. It is up to readers, as the designer.

An example will be given later on.

3.2.7.3. *Constraints imposed on antenna design*

In the case of the example of a given application, in general, the designer has a mechanical bulk with a maximum external surface area of x mm^2. Then, following discussions with the subcontractor making the printed circuit board, the designer learns that the manufacturer's equipment requires a minimum gap between the tracks of 0.2 mm (200 µm).

Additionally:

– A standard application for an integrated circuit typically recommends an inductance value between 0.4 and 4 µH. Indeed, at 13.56 MHz, the maximum inductance of the NFC device is relatively restricted to around 10 µH, because of the value of the minimum reproducible industrial physical capacity (~10–15 pF are typically deemed sufficient values to prevent all parasitic capacities, parasites dispersions and a self-resonance > 35 MHz) and its minimum value at ~300 nH by the small dimensions of coils that can be made on surfaces such as µSIM cards.

– Also, in principle, for application reasons (minimization of radiated electrical fields and/or simultaneous use of a screen), the antenna should have a mid-point for ease of implementation and manufacture of the printed circuit. It should preferably have an even number of turns.

– In order to produce the maximum magnetic field (which, at the same level of current, is proportional to the number of turns), we have opted for an antenna with as high a number of turns as possible (N of 1–6 turns), and which therefore has as high an inductance value as possible.

– However, NFC is, by definition, an application where the initiator and target sometimes need to be very close to one another. In order to reduce the coefficient and coupling index of the system, it is preferable to decrease the number of turns of the antenna to minimize possible detuning (initiator and/or target), so as to preserve the appropriate working distances. The reduction in performance due to this decrease in number of turns is hardly noticeable when the number of turns N is between 1 and 4.

Why have all these specifications been given? We are about to go on and discuss magnetic flux.

We suppose that we can work at 13.56 MHz and that:

– the P_max that the integrated circuit can deliver is known;

– the whole system works with "matched" impedances;

– the value of Q is known because of the ISO, which gives the value $R = L\omega/Q$.

$$I\_ant = \sqrt{[(P \times Q)/(L \times \omega)]}$$

The flux produced is equal to:

$$\Phi = L\ I\_ant$$

$$= L \sqrt{[(P \times Q)/(L \times \omega)]}$$

$$\Phi = \sqrt{(L)} \times \sqrt{(PQ/\omega)}$$

Also:

$$\Phi = B\ S$$

$$= \mu\ H\ S$$

Hence:

$$\mu\ H\ S = \sqrt{(L)} \times \sqrt{(PQ/\omega)}$$

For an antenna of a given shape, and therefore known mechanical dimensions, its total surface area "S" is equal to "N s", so therefore:

$$H = \sqrt{L}\ \frac{\sqrt{(PQ/\omega)}}{\mu\ N\ s}$$

We will also show that the value of the antenna's inductance L is equal to the product Lo N^p, with Lo being the inductance per turn, and very frequently an exponent p = 1.8–2, which gives us:

$$H = \sqrt{(Lo\ N^p)}\ \frac{\sqrt{(PQ/\omega)}}{N\ \mu\ s}$$

The example of calculations given in Table 3.6 summarizes and clearly indicates the maximum value of the magnetic field H_0_max that can be delivered by a chosen, known integrated circuit in the case of a given antenna (format, dimensions and electrical characteristics). If we know the value of that maximum magnetic field produced, it is easy to look at Table 3.5 to determine the maximum theoretical distance that our specific application can cover (except if we encounter environmental problems – see later on).

NOTE.– In addition, the right-hand column recaps the same calculations as used for a reference PCD in the conformity/test standard ISO 10373-6.

EXAMPLE.–

Mechanical – Spire				PCD ISO
Format			*Rectangular*	*Circular*
Width	A	mm	10	
Length	B	mm	50	
Section	s = a × b	mm²	500	17662.50
Electrical – Antenna				
Inductance	L see calculations in the Excel spreadsheet	µH	3.85	2.31
Number of turns	N		7	2.00
Exponent chosen	P		1.8	1.80
	N^p		33.20	3.48
Inductance per turn	Lo = L / N^p	µH	0.12	0.66
Integrated circuit used		NXP	CL RC 663	CL RC 663
	V_supply	V_dc	3.3	3.30
	V_out differential square	V_pp	6.60	6.60
	V_out_differential_rms_sine_equi	V_rms	1.49	1.49
	R_ic_out_differential	Ω	35	35.00
	P max matched load	mW	63.13	63.13
Application			xxx	PCD ISO
Frequency	F	MHz	13.56	13.56
	ω	10^6 rd/s	85.156	85.16
Magnetic permeability	µ	10^-7	12.56	12.56
	Q in load, in accordance with ISO 14443		30	30.00
	$\sqrt{(Lo\ N^p)} = \sqrt{L}$	10^-3	1.96	1.52
	H_0_max delivered by the circuit	A/m	66.567	5.109

Table 3.6. *Maximum magnetic field H_0_max that can be delivered by a chosen integrated circuit*

We have now fully defined the context of our study. Let us now give a brief recap of the theory of circular and rectangular antennas.

3.2.8. *Simple antenna*

For NFC systems working with inductive coupling at the frequency of 13.56 MHz, conventional antennas for the most widespread initiators (and also for targets/tags – see Chapter 6) for tiny and short distances are generally flat coils with a circular, square or rectangular shape.

3.2.8.1. *Flat circular coil*

Figure 3.23 shows the general shape of such antennas.

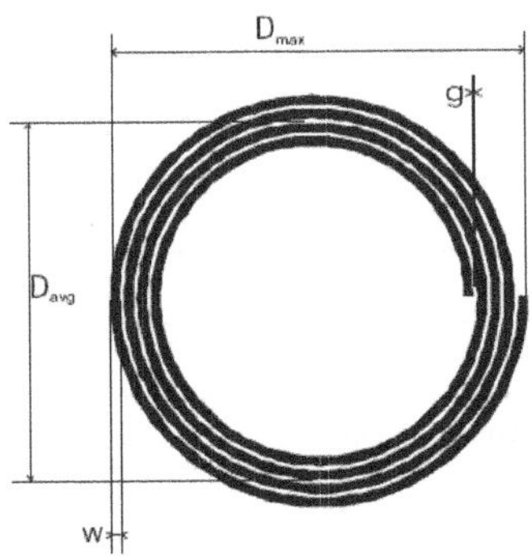

Figure 3.23. *Flat circular coil*

With a few second-order approximations, we can show that the equation governing the value of their inductances as a function of their number of turns is given by:

$$L = 2\ l\ [\ \ln (l/D)\ -\ k\ ] \times (N\text{^}p)$$

$$L = Lo \times N\text{^}p$$

Taking account of the coil's mechanical parameters, and setting:

Average diameter $\qquad D\_aver = Do - N(g + w)$

Circumference of a spire $l = \pi . D\_aver$

$$d = 2 (w + t)/\pi$$

The equation giving the value of the inductance for this type of setup is shown below:

$$L = 2 (l) [\ln (l/D) - 1.07] \times (N^2)$$

where l and D are expressed in cm, and L in nH.

In these equations:

L total inductance of the coil (in nH);

Lo inductance of a turn;

l circumference (= 2 $\pi \times r$) (in cm);

D diameter of the wire or width of the path of the conductor if the antenna is a printed circuit board (in cm);

N number of turns;

k correction factor, whose value depends on the antenna's geometry,

= 1.07–1.16 for circular antennas

(k is normally very close to ln (π) = 1.144).

In fact, the factor "k" represents the Neperian logarithm of a constant "C" whose value is equal to "c × π". The coefficient "c" represents the corrective term in view of the specific shape/geometry of the coil, in comparison to a circular coil with radius "r" much greater than the diameter "D" of the wire used. The value of "c" thus floats around 1 (between 0.8 and 1.5) depending on the geometric shape:

$$\ln (l/CD) = \ln (l/D) - \ln C, \text{ where } \ln C = \ln (c \times \pi) = \ln c + \ln (\pi)$$

$$\ln C = \ln c + 1.144$$

By setting: ln C = k, we obtain: $\ln (l/CD) = \ln (l/D) - k$

EXAMPLES.–

– For square antennas, $k = 1.47 = \ln C = \ln c + 1.144$

so $\ln c = 0.326 \rightarrow c = 1.385$

– For rectangular antennas, $k = 1.04 = \ln C = \ln c + 1.144$

so $\ln c = -0.104 \rightarrow c = 0.901$

p is an exponent whose value depends on the coil technology in the antenna, and therefore on the relative proximity of the turn to one another.

Indeed, because of the actual mechanical construction of an antenna coil, the spire-to-spire coupling will never be exactly equal to 100%, and consequently the value N^2 typically found in the theoretical mathematical relation is physically illusory. Hence, we are led to redefine the value of the exponent "p", which gives:

p	= 1.8-1.9	for wire-wound antennas
p	= 1.7-1.85	for etched antennas
p	= 1.5-1.75	for printed antennas
ln	Neperian logarithm	

3.2.8.2. *Flat rectangular coil*

In view of the mechanical parameters of the coil, the exact equation giving the value of the inductances for rectangular coils/antennas is shown below:

$$L1: \frac{UO}{\pi} \times \left[a.\ln\left[\frac{2 \cdot a \cdot b}{11 \cdot \left(a + \sqrt{a^2 + b^2}\right)}\right] + b.\ln\left[\frac{2 \cdot a \cdot b}{11 \times \left(b + \sqrt{a^2 + b^2}\right)}\right] - 2 \cdot \left(a + b - \sqrt{a^2 + b^2}\right) + \frac{a+b}{4} \right] \cdot N1^{1.85}$$

This formula was developed by Franz Amtmann (NXP Semiconductors), Sébastien Rieubon, Pascal Roux (Xerox ex. ACS)

Figure 3.24 shows the general shape of such antennas.

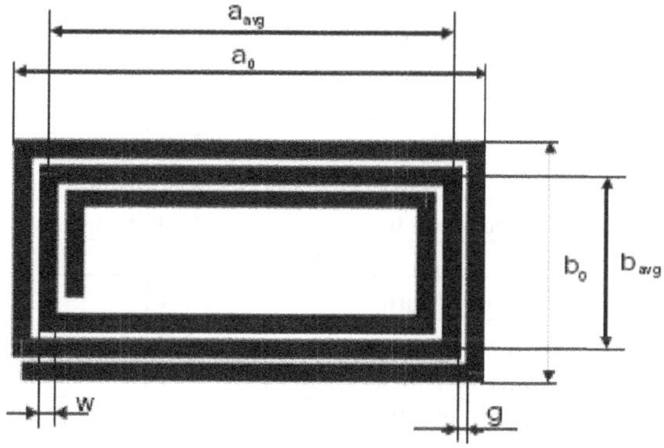

Figure 3.24. *Flat rectangular coil*

By setting:

Average length	a_aver	= ao – N(g + w)
Average width	b_aver	= bo – N(g + w)
	l	= π . D_aver
	d	= 2 (w + t)/π

$$x1 = a\_aver \,.\, \ln\left[\frac{2 \,.\, a\_aver \,.\, b\_aver}{d\left[a\_aver + \sqrt{(a\_aver^2 + b\_aver^2)}\right]}\right.$$

$$x2 = b\_aver \,.\, \ln\left[\frac{2 \,.\, a\_aver \,.\, b\_aver}{d\left[b\_aver + \sqrt{(a\_aver^2 + b\_aver^2)}\right]}\right.$$

$$x3 = 2\left[(a\_aver + b\_aver) - \sqrt{(a\_aver^2 + b\_aver^2)}\right]$$

$$x4 = (a\_aver + b\_aver)/4$$

we find:

$$L = \frac{\mu\_0}{\pi}\,(x1 + x2 - x3 + x4)\,N^p$$

By applying the above formulae for circular and/or rectangular antennas, we run a specific "Excel spreadsheet" optimization tool multiple times in order to gain an initial set of "coherent" results on the mechanical dimensions of the coil, the value of the inductance and the minimum surface area, to create a model with a view to industrialization.

Example of results for a rectangular antenna coil measuring 40 × 20 mm.

Overall length	40	mm
Overall width	20	mm
Track thickness	35	μm
Track width	300	μm
Equi. diameter of track	213	μm
Gap between tracks	500	μm
Number of turns	4	
Turn exponent, p	1.7	
Average length	36.8	mm
Average width	16.8	mm
Average coil area	618.24	mm^2
Inductance	0.95	μH

Table 3.7. *"Coherent" results on the mechanical dimensions, inductance value and minimum surface area*

Although the values of the copper thickness and track width used to compile these tables are well-established standards in the profession, it will still be necessary for designers to ask their printed circuit supplier(s) whether this is compatible with their manufacturing tools – at a given cost, of course.

3.2.8.3. *A few remarks on the content of the table*

– *The exponent "p"* (in Table 3.7) has been chosen arbitrarily while we await the final choice of technology.

– Antenna resistance.

– Ohmic resistance in a DC regime: we can calculate the value of the antenna's DC resistance:

R_ant_dc = (ρ N_ant l)/s   ρ = resistivity of the wire used

The resistivities ρ of copper Cu and aluminum Al are:

ρ Cu = 1.7 10^-8Ωm and ρ Al = 3 10^–8Ωm
N_ant = number of turn
l = average turn length
s = rms section of the wire

This value of R_ant_dc is practically always lower than that which is need to match the quality coefficient. Consequently, we are able to match it externally to the final value desired for the application (approximately 20–35).

– Resistance at 13.56 MHz due the skin effect: the skin depth δ is essentially the width/depth of the area in which the current is concentrated within a conductor. At a given frequency, it enables us to calculate the rms resistance. Its value is given by the following equation:

$$\delta = \sqrt{\frac{2}{\omega.\mu.\sigma}} = \sqrt{\frac{2.\rho}{\omega.\mu}} = \frac{1}{\sqrt{\sigma.\mu.\pi.f}}$$

where:

– δ: skin depth in meters [m];

– ω: pulsation in radian per second [rad/s] (ω=2.π.f);

– f: frequency of the current in hertz [Hz];

– μ: magnetic permeability in Henry per meter [H/m];

– ρ: resistivity in ohm per meter [Ω.m] (ρ=1/σ).

EXAMPLES.– For a copper conductor, we find the following values for the skin depths δ.

Frequency	Δ
100 kHz	0.21 mm
1 MHz	66 μm
10 MHz	21 μm
13.56 MHz	17.6 μm

Table 3.8. *Values of the skin depths δ*

For example, on a broad track with width 1 mm and depth 35 μm, it can be considered that the current is not entirely uniform throughout the cross-section of the wire, and that the skin effect may be a concern. At a given frequency, for a circular conductor whose (equivalent) diameter is significantly greater than δ, we can calculate the rms resistance, considering that only the outer part, with depth δ, contributes to conduction. For example, for a cylindrical conductor with radius R, we would have a useable section of:

$$S_u = \pi.(R^2 - (R - \delta)^2)$$

EXAMPLE.–

section	track	= 1,000 μm × 35 μm
		= 35,000 μm^2
in DC	R_dc_/_meter	= (1.7 10^-8)/(35000 10^-12)
		= 0.485 ohm for a meter
in HF	r_equi	= 105 μm
	s_useable	= 3.14 (105^2 - (105 – 17)^2)
	s_useable	= 10302 10^-12m^2
	R_hf_/_meter	= (1.7 10^-8)/(10302 10^-12)
		= 1.65 ohm for one meter

3.2.9. *Matching circuit for the impedance of the antenna*

As we now know the couple of values L, R of the antenna, we merely need to bring the antenna's impedance into line with the value of Rp (in case of application with an EMC filter) or Rs (without an EMC filter), and the optimal matching of power to the integrated circuit will (almost) be achieved. For this purpose, a purely reactive (and therefore non-dissipative) setup, matching the impedance between the EMC filter and the antenna, known as a "capacitance bridge", is generally used. The simple diagram of this setup, well known to people in the field of RF, is presented in a single-ended configuration in Figure 3.25.

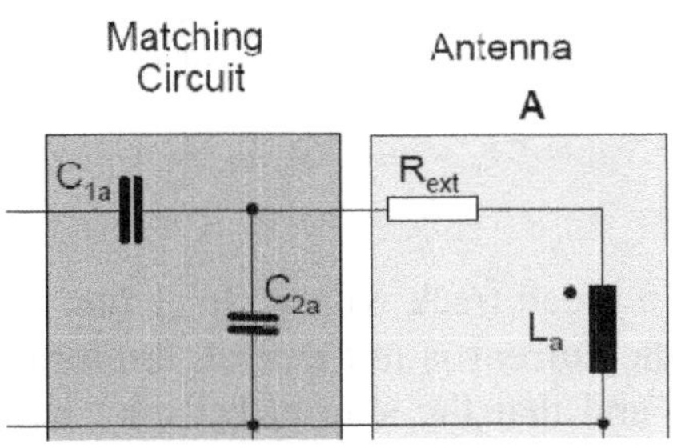

Figure 3.25. *Matching circuit presented in a single-ended configuration*

3.2.9.1. *Determination of the impedance-matching capacitance bridge*

The complex input impedance equation Z_in for the whole of the network thus constructed has the value:

$$Z\_in = \frac{1}{j\,C1\,\omega} + \frac{\dfrac{1}{j\,C2\,\omega} \times (R1s + j\,L1s\,\omega)}{\dfrac{1}{j\,C2\,\omega} + (R1s + j\,L1s\,\omega)} = (a + j\,b)$$

Once again, at the operating frequency of the whole circuit, in order for this impedance matching to be fully satisfied, the following conditions must be fulfilled at the same time:

– the *input impedance Z_in must be purely real,* meaning that its imaginary part is null, so the value of the impedance phase is equal to 0 degrees;

– the *modulus of the real part of Z_in – "R_in" – must be equal to the value R_out desired* by the application designer (for instance, to R_out of the amplifier or EMC filtering circuit so as to produce the best possible matching of the transferred power).

In other words, we must be able to write the above equation in the form:

$$Z\_in = (a + j\,b) = R\_in$$

This means:

– first a = R_in;

– second b = 0.

The complete development of the equation for Z_in is lengthy (and goes beyond the remit of this book. Besides, readers have observed the process at the beginning of this chapter). By taking account of the last two conditions for "a" and "b", we can determine unique values for the capacities C1 and C2 as a function of the other elements. Sparing the long-winded intermediary equations, we obtain the following two conditions, *with no approximation whatsoever* (approximations are too often given in numerous published works):

$$C1 = \frac{1}{\omega} \times \frac{\sqrt{R1s}}{\sqrt{[(R\_in) \times (R1s^2 - (R1s \times R\_in) + (L1s^2 \times \omega^2))]}}$$

and $$C2 = \frac{(R\_in - R1s)}{(L1s \times \omega^2 \times R\_in) + (R1s/C1)}$$

NOTES.– In the applications of interest to us here, because the value of ω is always high, and often – though not always – the term $(L1s^2 \times \omega^2)$ is much greater than $(R1s^2 - (R1s \times R\_in))$ and, as an initial approximation, the value of C1 is basically equal to:

$$C1 = \frac{1}{(L1s \times \omega^2)} \times \sqrt{\frac{R1s}{R\_in}}$$

By substituting the value of C1 into the equation of C2, obviously, we can find the value of C2.

If we now consider the same setup represented, after transformation, in its equivalent parallel form, we know that we can write:

$$R1p = Q1^2 \times R1s \quad \text{where} \quad Q1 = \frac{L1s \times \omega}{R1s}$$

Hence:

$$R1p = (L1s^2 \times \omega^2)/R1s$$

If we substitute this value back into the above equation, we find:

$$C1 \approx \frac{1}{\omega \times \sqrt{(R\_in \times R1p)}}$$

Due to the C2 value and because the value of R1s is very small and the value of $(L1s \times \omega^2)$ is large, and also $L1s \approx L1p$, we can write:

$$C2 \approx \frac{1}{L1p \times \omega^2} - C1 - Cp$$

NOTE.– It is not difficult to switch from a diagram of a single-ended matching setup to one of a differential matching setup: we simply need to double the values of the capacities C1 and C2. Thus, the new values are C1', C1'' and C2', C2'', which are both, two by two, respectively, in a serial setup.

N.B.– *If the EMC filter is not strictly tuned to the carrier* but instead to a value slightly higher than that of the carrier frequency, then its output impedance is a complex value equal to (a + j b) (rather than simple to R) – see Figure 3.26.

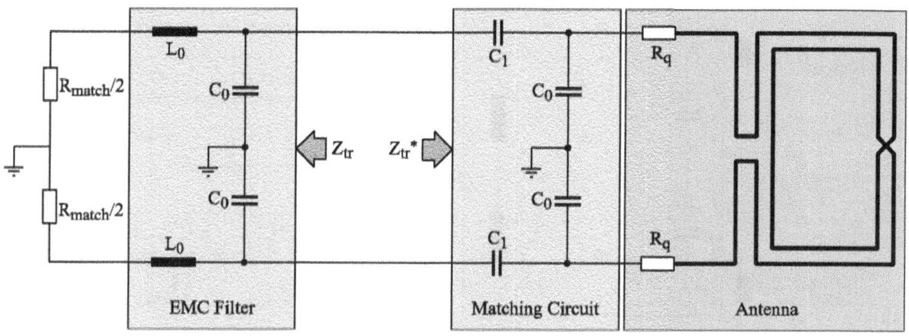

Figure 3.26. *Diagram of a matching setup in differential mode*

The matching circuit, made up of the "capacitance bridge + antenna", must behave as an impedance with the conjugate value Z* = R – jX. In this case, the values of C1 and C2, respectively, become:

$$C_1 \approx \frac{1}{\omega.\left(\sqrt{\frac{R_{tr}.R_{pa}}{4}}+\frac{X_{tr}}{2}\right)}$$

$$C_2 \approx \frac{1}{\omega^2.\frac{L_{pa}}{2}} - \frac{1}{\omega\sqrt{\frac{R_{tr}.R_{pa}}{4}}} - 2.C_{pa}$$

3.2.9.2. *Calculating the matching elements of the antenna*

Here, we can calculate the values by a quick operation in Excel (the NXP document). Note that the choice of value of the inductance L for the initiator antenna must give the values of C1a and C1b that ensure the setup is reproducible (meaning that these values cannot be ridiculously low).

To recap, Figure 3.27 shows the value ranges recommended, for example, by NXP when the desired matching is 700 Ω.

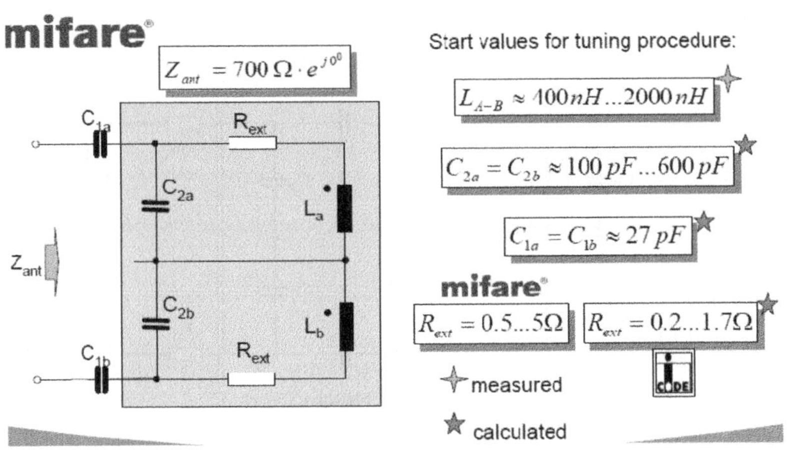

Figure 3.27. *Values recommended by NXP for 700 Ω matching*

As we can see in this figure, the main aim of the value of the external resistance R_ext is to match the value of the global quality coefficient Q of the setup to the application. This is directly linked to, and is a function of, the standard ISO 14443 (in NFC IP1 and IP2) or ISO 15693 (in NFC IP2) (the bandwidth necessary to communication timings, etc.).

To recap, in general, the maximum useable values are:

ISO 14443 Q_max = 35

ISO 15693 Q_max = 100

which gives us R1 total = R copper + R_ext (see Figure 3.28), and also the thickness of copper, length of the tracks and number of turns.

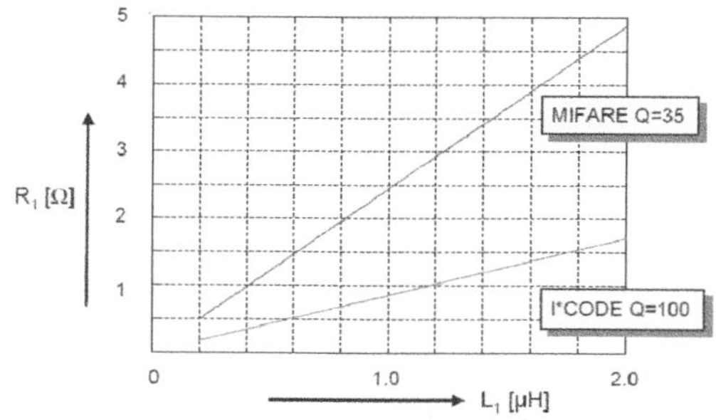

Figure 3.28. *R = f(L)*

3.2.10. *Calculating the current in the antenna coil of the initiator*

Remember that the impedance of the "matching + antenna" network C1, C2, L1, R1 is:

$$Z\_in = \frac{1}{j\,C1\;\omega} + \frac{\dfrac{1}{j\,C2\;\omega} \times (R1s + j\,L1s\;\omega)}{\dfrac{1}{j\,C2\;\omega} + (R1s + j\,L1s\;\omega)} = (a + j\,b)$$

In the conditions of tuning, there is maximum power transfer when R_int = "a" and the current given by the generator is then:

$$\begin{aligned} V &= (R\_int + a)\,I \\ &= 2\,R\_int\;I \end{aligned}$$

The voltage across the terminals of the antenna (L1, R1) is equal to V_eq_rms = | Z_eq | × I_rms, where:

$$Z\_eq = \frac{\dfrac{1}{j\,C2\;\omega} \times (R1s + j\,L1s\;\omega)}{\dfrac{1}{j\,C2\;\omega} + (R1s + j\,L1s\;\omega)}$$

As we know, when the circuits are tuned, we will have (see above):

$$C2 = \frac{1}{L1p \times \omega^2} - C1 - Cp$$

$L1p = L1s$

$C\_e = C1 + Cp$

$$C2 = \frac{1}{L1s \times \omega^2} - C\_e$$

$$Z\_eq = \frac{(R1s + j\,L1s\,\omega)}{(L1s\,C\_e)\,\omega^2 + j\left(\frac{R1s}{L1s\,\omega} - R1s\,C\_e\,\omega\right)}$$

Consider the modulus of Z_eq:

$$|Z\_eq| = \frac{|(R1s + j\,L1s\,\omega)|}{\left|(L1s\,C\_e)\,\omega^2 + j\left(\frac{R1s}{L1s\,\omega} - R1s\,C\_e\,\omega\right)\right|}$$

Hence, $V\_eq\_rms = |Z\_eq| \times I\_rms$

Knowing V_eq_rms, we derive the current I_Ls_rms in the antenna:

$$I\_Ls\_rms = \frac{|Z\_eq| \times I\ rms}{|(R1s + j\,L1s\,\omega)|}$$

$$I\_Ls\_rms = \frac{I\_rms}{\left|(L1s\,C\_e)\,\omega^2 + j\left(\frac{R1s}{L1s\,\omega} - R1s\,C\_e\,\omega\right)\right|}$$

Supposing that Cp is very small, Ce = C1:

$$C1 = \frac{1}{\omega\ \sqrt{(R\_in\ Q1^2\ R1s)}}$$

$$C1 = \frac{1}{\omega\, Q1 \;\sqrt{(R\_in\; R1s)}}$$

and by substituting back in, we find:

$$I\_Ls\_rms = \frac{I\_rms}{\left| \frac{1}{\sqrt{(R\_in/R1s)}} + j\,\frac{1}{Q1}\left(1 - \frac{1}{\sqrt{(R\_in/R1s)}}\right) \right|}$$

NOTE 3.1.– In practice, Q1 is often large, so the value of 1/Q1 is small and the next parenthesis has a small value, so the imaginary term in "j" is very small, and often I_Ls can be approximated as:

$$I\_Ls\_rms \sim= I\_rms \times \sqrt{(R\_in/R1s)}$$

and independent of the value of L1s, but dependent upon the value of R1s by way of the value of the quality coefficient Q1 of the circuit R1s, L1, C.

NOTE 3.2.– THREE STARS! Let us look again at the wonderful equation that we have just obtained, and square it:

$$I\_Ls\_rms^2 \sim= I\_rms^2 \times (R\_in/R1s)$$

or indeed:

$$R1s\; I\_Ls\_rms^2 \sim= R\_in\; I\_rms^2$$

$$P\_watts\_output = P\_watts\_input$$

...which has already been known for a very long time, and is nothing out of the ordinary, because filters and matching circuits are only reactive, and therefore non-dissipative, and thus with that equality of the power levels between the input and output, we can immediately calculate the current in the inductance that is the antenna. It may seem overly fastidious to perform all these calculations just for that ... but now readers have all the necessary tools at their disposal to calculate reactive elements and to fill in the tables and understand what they are about.

3.2.11. *Summary and examples*

3.2.11.1. *Application without an EMC filter*

To conclude this chapter, let us substitute the value of I_rms = (V_batt/R_int)/π √2 into the equation of I_Ls_rms:

$$I\_Ls\_rms \sim= \; I\_rms \times \sqrt{(R\_in/R1s)}$$

$$I\_Ls\_rms \sim= \; ((V\_batt/R\_int)\sqrt{(R\_in/R1s)})/\pi\sqrt{2}$$

which, ultimately, gives us the following very useful equation:

$$I\_Ls\_rms \sim= \; (V\_batt/\sqrt{(R1s\,R\_in)})/(\pi \times \sqrt{2})$$

independent of the value of L1s, but dependent upon the value of R1s by the value of the quality coefficient Q of the LC circuit.

EXAMPLES.– With a total inductance of twice 1 μH in a serial circuit = 2 μH (see Table 3.9).

Circuit Inside Secure	V_batt = 3.3 V		
Setup Single-ended	R_in = 12.5Ω		
	L1s in μH	R1s (= L1s ω / Q) in Ω	I_Ls_rms in mA_rms
ISO 14443 Q = 35	1	2.45	
Proximity	1.5	3.76	
(H = from 1.5 to 7.5A/m)	2	4.89	94
ISO 15693 Q = 100	1	0.85	
Vicinity	1.5	1.28	
(H = from 150mA/m to 5A/m)	2	1.71	

Table 3.9. *Application without an EMC filter*

NOTE.– We could, in fact, have established this relation very easily by saying that all the reactive elements in the impedance matching circuits consume no power, and therefore when the impedance is matched, all the power from the generator passes into the load, which is R1s, giving us the following simple equality:

$$P\ gene = V\_batt^2/(2\,\pi^2\,R\_int) = R1s\,(I\_Ls\_rms)^2 = P\ load$$

$$I\_Ls\_rms^2 = V\_batt^2/(2\ \pi^2\ R\_int\ R1s)$$

$$I\_Ls\_rms = (V\_batt/\pi\ \sqrt{2})\ \times\ 1/\sqrt{(R1s\ R\_int)}$$

EXAMPLE of NXP circuit in differential mode.–

$$I\_Ls\_rms = (6.6/\pi\ \sqrt{2})\ \times\ (1/\sqrt{(5.11 \times 50)}) = 94mA.$$

3.2.11.2. *Example of application with an EMC filter and R_in = 12.5 ohms*

To conclude all these calculations, the summary table is below (Table 3.10).

f	=		13.56	MHz	Figure
ω	= 2 π f		85.72	10^6 rd/s	
Characteristics of the setup and integrated circuit					
setup	= Single-ended				
V_batt	=	V of power supply to IC	3.3	V_rms	
R_ic_out	= data sheet	Internal R of the IC	12.5	ohms	
I_ic_p	= (4/π × (V_batt)/2 / (R_in + R_load))		84	mA_p	
I_ic_match_ rms	= (4/π × (V_batt)/2 / (R_in + R_load)) / √2		60	mA_rms	
Hypothesis of EMC "filtering" matched to R_in					
Chosen L	=	User's choice	200	nH	
Calculated Cp	= See EMC filter calculations in table		429.6	pF	
Calculated Rp	= See EMC filter calculations in table		41	Ω	
"Antenna" hypothesis					
N	=	Number of turns	4	turns	
Shape	= rectangular		4 × 2	cm	
r_equi	= equivalent radius of circular surface.		1.9	cm	
L1s	= see specs in Excel table	L of the antenna	2	μH	
R_bob	=	R of the antenna	1	Ω	
Q1	= L1s ω / R1s	Imposed by application ISO 14443	35		
R1s	= L1s ω / Q1		4.89	Ω	
R_ext	= R1s – R_bob	Additional R	3.89	Ω	
R1p	= Q1² R1s		5.99	kΩ	
Impedance matching circuit with Z_in = Rp equivalent					
C1	= 1/(ω × RMS (Rp × R1p)		23.5	pF	
C_p	= Parasitic capacitance	Estimated value	0	pF	

C_e	= C1 + Cp		23.5	pF	
C2	= (1/(L1s ω²)) - C_e	68 – 23.5	44.5	pF	
Calculation of I_Ls_rms					
R_in/R1s	= 41 / 4.89		8.2		
√(R_in/R1s)	=√		2.86		
I_Rp_rms	= See table of EMC filter calculation		32.8	mA_rms	
I_L1s_rms	= √ (R_in / R1s) × I_Rp_rms = 2.86 × 32.8	I in the antenna	93.9	mA_rms	QED
I_L1s_p	= I_L1s_p × √2		132.9	mA_p	
Calculation of V_L1s_rms					
V_L1s_rms	= L1s × ω × I_L1s_rms = Q × R1s × I_L1s_rms	V at the antenna terminals	16	V_rms	
V_L1s_p	= V_L1s_rms × √2	V at the antenna terminals	22.64	V_p	
Calculation of the power dissipated in the resistive antenna circuit					
P_rms	= R1s × (I_L1s_rms)²		44	mW_rms	QED

Table 3.10. *Example of application with an EMC filter and R_in = 12.5 ohms*

3.2.12. *Simulations*

3.2.12.1. *Application with an EMC filter*

As is shown by Figures 3.29–3.32, the simulations match the previous calculated values.

Let us be serious and state simply that, in this chapter, we have given a detailed picture of where these values and oscillograms come from, and that when dealing with new projects, readers will be able to very quickly choose, define and optimize the solutions, which was really the very substance of this chapter.

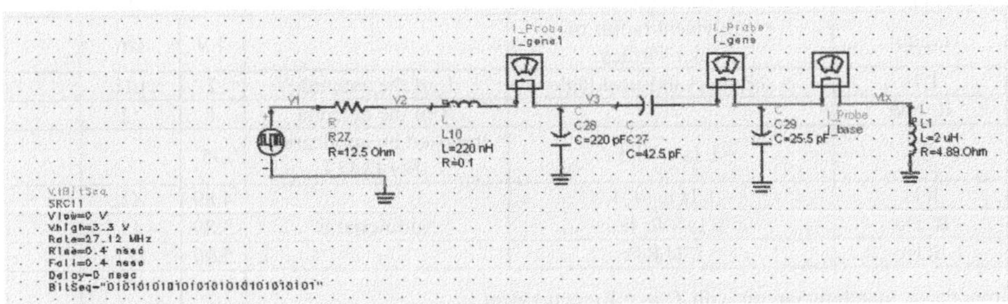

Figure 3.29. *Simulation diagram – application with an EMC filter. For a color version of the figure, see www.iste.co.uk/paret/antenna.zip*

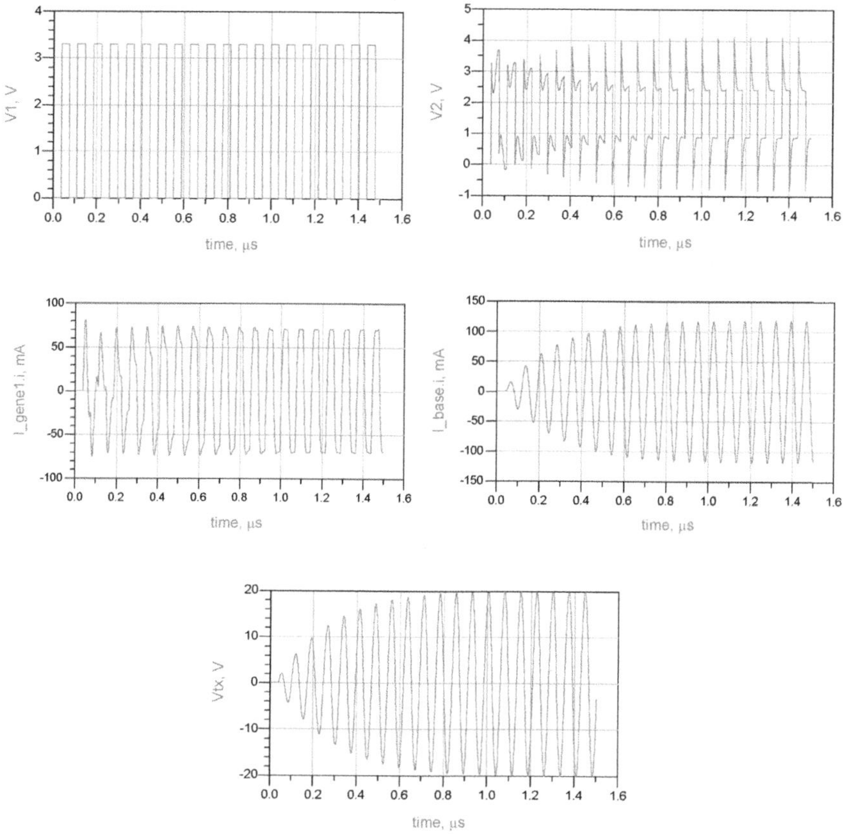

Figure 3.30. *Simulation results – application with EMC filter*

3.2.12.2. *Final example of simulation with second filtering LC circuit*

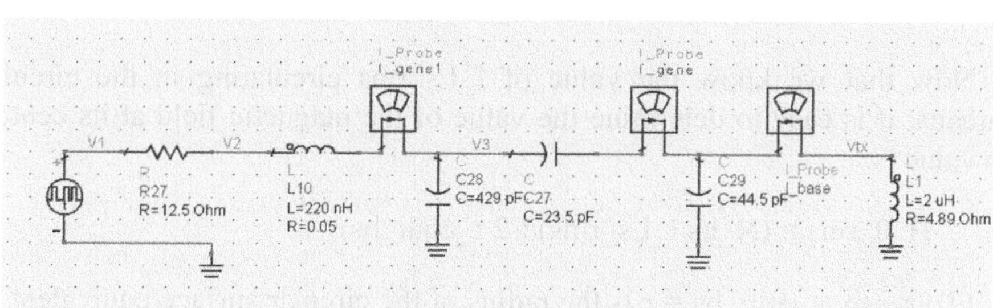

Figure 3.31. *Simulation diagram – application with an EMC filter . For a color version of the figure, see www.iste.co.uk/paret/antenna.zip*

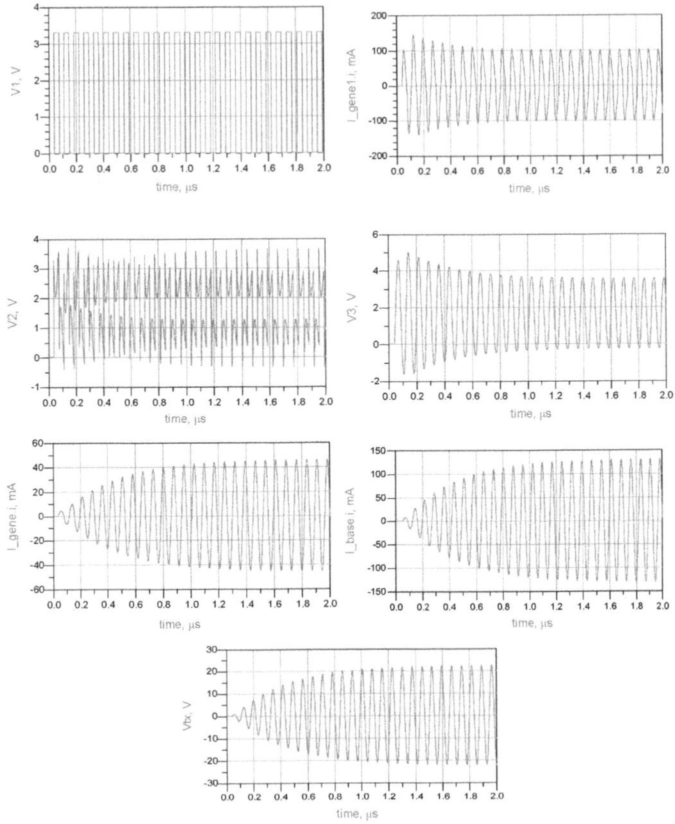

Figure 3.32. *Simulation results – application with an EMC filter*

3.2.13. *Value of the field H radiated by the antenna*

Now that we know the value of I_Ls_rms circulating in the circular antenna, it is easy to determine the value of the magnetic field at its center. Its value is:

H_0_rms = (N_bs I_Ls_rms)/(2 r_equi_bs)

(To recap, r_equi_bs = r is the radius of the circular surface equivalent to the total area of the real antenna).

3.2.14. *Calculation and value of the working distance*

We can, therefore, calculate the relation "field/number of spires/r_equi_bs/distance d", which is summarized by Table 3.11, in which a horizontal row is read as follows:

1) we continue with the example, and knowing that the current in the antenna I_Ls has the value of 94 mA_rms (see all the above sections), at the desired field intensity H_0_rms = x, this leads to a value of N/r = y, because 2 H_0/I_Ls = N/r;

2) knowing the equivalent radius of the initiator's coil "r", via the value obtained above for N/r, we deduce the number of turns needed in order to create the desired field. (It is to be hoped that the number of turns and the shape of the coil of the antenna gives the desired inductance, but we need to check it for the sake of caution);

3) we suppose that we have an NFC tag just at the boundary of ISO 14443/18092, which therefore requires a field H_d = 1.5 A/m in order to work. Thus, we can calculate the ratio between the field H_0 supplied in the antenna's plane to d = 0 and H_d, at a distance which is as yet unknown;

4) if we know the physical/mathematical equation which links the decrease in the axis of the magnetic field of the antenna with the distance d, that ratio H_d/H_0 leads to the knowledge (determination) of the ratio d/r;

5) finally, when we know r, we can deduce the working distance d in those operating conditions.

3.3. Maximum quality coefficient Q of the initiator antenna

Many times in this chapter, we have mentioned the theoretical values of the quality coefficient (Q). This involves knowing a number of details and going back theoretically to the value of Q, which has two influences:

– on the value of the bandwidth, and thus on the achievable bit rate;

– on the shape of the field cutoff – particularly when the bit stream is paused.

Without the presence of a tag in the field and with no loading effect whatsoever					
Hypotheses	I_Ls = 94 mA	r_equi = 1.5 cm			
ISO 14443			If we set H_d = 1.5A/m (ISO 14443)		
H_0_rms	N/r_equi	N_min	H_d/H_0	d/r	D
A/m	= 2 H_0/I_Ls	of spires			cm
1.5	32	0.48	1	0	0
3	63.82	0.95	0.5	0.7	
3.75	79.78	1.19	0.4	0.85	
4.5	95.74	1.43	0.33	1.2	
7.5	159.57	2.39	0.2	1.4	2.1
Hypotheses	I_Ls = 94 mA	r_equi = 1.5 cm			
ISO 15693			H_d = 150 mA/m (ISO 15693)		
H_0_rms	N/r_equi	N_min	H_d/H_0	d/r	D
A/m	= 2 H_0/ I_Ls	of spires			Cm
0.150	3.2		1	0	0
1.5	32				
2					
3	63.8	0.957			
5	106.38	1.6	0.03	3	4.5

Table 3.11. *Calculation and value of the working distance*

NFC uplink connections use carrier modulation of the type "amplitude-shift keying (ASK) m%". The maximum value of the overvoltage coefficient Q_max of the initiator antenna must be such that the bandwidth Bp_ant (at –3dB), equal to f0/Q, is at least capable of carrying all the frequencies contained in the spectrum of the signal modulating the carrier. In the worst-case scenario, the fundamental frequency (bit rate) of the square-wave signals (with cyclic ratio of 50% in the case of non-return-to-zero (NRZ) bit codes) of that digital data stream must, at least, be equal to half the bandwidth Bp_ant of the tuned circuit – i.e.:

$$2 \times \text{bit\_rate} = \text{Bp\_ant\_min},$$

or indeed

$$2 \times \text{bit\_rate} = \text{f0/Q\_max}$$

Hence

$$\text{Q\_max} = \text{f0} \times [1/(2 \times \text{bit\_rate})]$$

EXAMPLE.–

– carrier frequency: f0 = 13.56 MHz;

– digital flow of data transmitted by the initiator: bit_rate = 106 kb/s.

In this case, we find: Q_max = 64.

Unfortunately, for uplink, NFC ISO 18092 (or ISO 14443 type A), the signal modulating the carrier is far from having NRZ 50/50 bit coding, and it is modified Miller bit coding in which, for a bit duration of 9.44 µs, a carrier "pause" of duration Tp = 3µs is carried out. This pause Tp is equivalent to the transmission of a frequency whose period is 2Tp = 6 µs, which is an equivalent bit rate = (1/2Tp) = 166 kb/s.

Hence		Q_max = f0 × [1/(2 × 1/2Tp)]
	==>	Q_max = f0 × Tp
meaning that		Q_max = 40.68

This theoretical maximum value is not acceptable, because it is always necessary to secure the systems in all scenarios (tolerances, drifts – L, C, frequencies, etc.). Suffice it to say that, in the above example, we could use a Q_max_useable of around 35.

NOTE.– As we will see, generally, these values of Q_max_useable are not too difficult to obtain, because the technological designs of the antennas give specific Q_max (no-load) values that are far higher than those required for the applications, and it is easy to reduce these values using resistors arranged in a series, as we will see later on.

Thus, we have resolved the question of the maximum applicative value of the quality coefficient that the initiator antenna can exhibit.

3.3.1. *Q and cutoff of the field*

The quality coefficient Q is also one of the main parameters influencing the cutoff and recovery of the carrier during "pauses" in ASK 100% and the rapid variations in modulation in ASK m%. For this purpose, let us examine the pulsed response of the antenna circuit.

In an amplitude modulation, of type ASK 100% or m%, during a pulsed transition as a voltage step E is applied to a serial antenna circuit R, L, C is said to be "resonant damped". The equation of the value of current flowing through the circuit during the transition, written after the Laplace–Carson transform, is well known and is of the form:

$$I = \frac{E}{L} \times \frac{p}{(\omega 0^2 + 2\,\Delta\,p + p^2)} = \frac{E}{L} \times \frac{p}{\omega 0^2 \left(1 + \frac{2\,\Delta\,p}{\omega 0^2} + \frac{p^2}{\omega 0^2}\right)}$$

In summary, with 2Δ = R/L, and setting:

f0	= carrier frequency	
R	< RMS (4L/C)	
ω0	= 1/RMS LC	==> L C $\omega 0^2 = 1$
ω0	= 2 π f0	
T0	= 1/f0	

After inverse transformation, the resulting signal is expressed by a variation in voltage (decrease or increase depending on the polarity of the variation), exponential in amplitude of the oscillations with period T0. With a few second-order approximations when excitation is stopped or resumed, in ASK 100%, the mathematical expression of the variations in voltage at the terminals of the inductance in the oscillating circuit is presented in the form:

v (t) = Vmax × cos ω t × [e^(- Δ t)]

and "[e^(- ΔT)]" is called the "logarithmic decrement".

Thus, this variation is a cosinusoid with an exponential envelope whose time constant is equal to q = 1/Δ = 2L/R. The equation can then be written as:

v(t) = Vmax × cos ω t × [e^(- t/q)]

Figure 3.33 shows an example.

In the knowledge that, for the initiator antenna, Q1= L1s ω0/R1s, we can also write that:

	Q1	= 2L (2 π × f0)/2R = 2L (π × f0)/R
but	q	= 1/Δ = 2L/R

$$q = Q1/(\pi \times f0)$$

$$= (Q1/\pi) \times T0$$

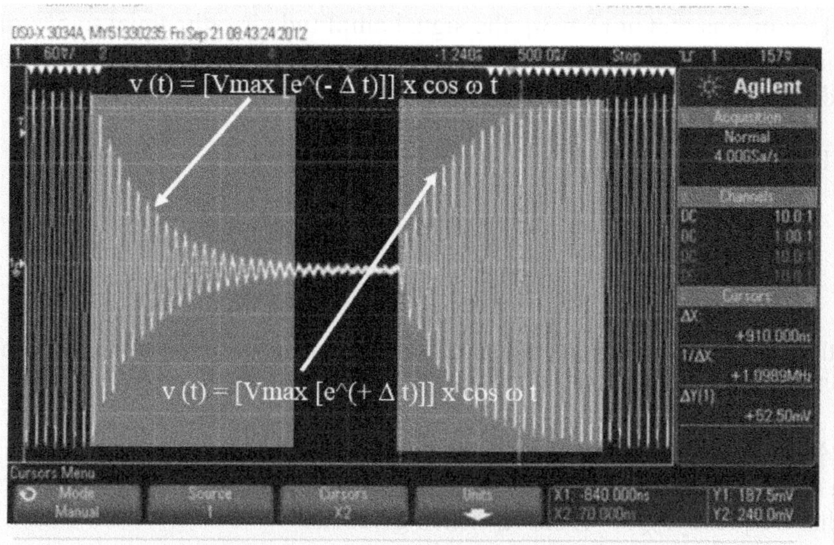

Figure 3.33. *Cosinusoid with an exponential envelope*

Thus, we have the relation between q and Q1:

$$\frac{q}{T0} = \frac{Q1}{\pi}$$

Let us, for example, calculate the values of the voltages at time t1= q and t3 = 3q.

– for t1 = q $\quad v(q) = Vmax \times e^{\wedge}{-1}$

so $\quad v(q) = Vmax/2.718$

so $\quad v(q) = 36\%$ of Vmax

– for t3 = 3 q $\quad v(3\ q) = Vmax \times e^{\wedge}{-3}$

so $v(3\ q) = Vmax/20$

so $v(3\ q) = 5\%$ of Vmax

This means that at t3 = 3q, the signal amplitude is only equal to 5% of its initial maximum amplitude. If the initiator works by ASK 100% amplitude modulation:

That time t3 = T_cutoff at 3 q

$= 3 \times (1/a)$

$= 3 \times (2L/R)$

is, therefore, the time period needed for the carrier to be physically (nearly, at 5%) cutoff.

At the working frequency of the setup (f0, T0 = 1/f0, $\omega 0 = 2\ \pi$ f0), we can also encode T_cutoff_x% as a function of the quality factor of the circuit at that frequency, $Q1 = L\ \omega 0/R$, which gives us:

As $Q1 = 2L\ (2\ \pi \times\ f0)/2R = 2L\ (\pi \times\ f0)/R$

$T\_cutoff\_36\% = 1q = (Q1/\pi\ f0) = (Q1/\pi\ ) \times T0$

$T\_cutoff\_5\% = 3q = 3 \times (Q1/\pi\ f0) = 3\ (Q1/\pi\ ) \times T0$

3.3.2. *Decrease in the ISO field*

Figure 3.34 shows the example of the NFC standards IP1 and IP2 (ISO 18092 and 21481), which conform to ISO 14443 A.

$(t1 - t2)max = (40.5\text{-}7)/fc = \sim 42/fc$

$= 73.7ns \times 42 = 3\mu s$

$(t1 - t2)min = (28\text{–}7)/fc = 21/fc$

$= 73.7ns \times 21 = 1.5\mu s$

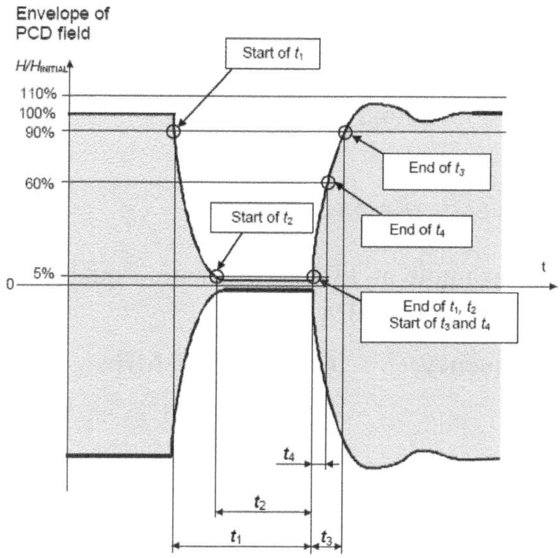

Parameter	Min	Max
t_1	$28/fc$	$40.5/fc$
t_2	$7/fc$	t_1
t_3	$1.5 \text{ x } t_4$	$16/fc$
t_4	0	$6/fc$

Figure 3.34. *Values to be respected according to NFC standards IP1 and IP2 (ISO 18092 and 21481)*

Consider an average time of (3 +1.5)/2 = ~2.2µs, which is the equivalent of ~ 31 carrier periods ... so we have the value of Q for NFC ISO 18092 and ISO 14443.

T_cutoff_5% $= 3\text{ q} \quad = 3 \times (\text{Q1}/\pi \text{ f0}) = 3\ (\text{Q1}/\pi) \times \text{T0}$

$3000 \text{ ns} = 3\ (\text{Q1}/\pi) \times 73.74 \text{ ns}$

$\text{Q1} = 42.58$

T_cutoff_5% $= 3\text{q} \quad = 3 \times (\text{Q1}/\pi \text{ f0}) = 3\ (\text{Q1}/\pi) \times \text{T0}$

(average) $2200 \text{ ns} = 3\ (\text{Q1}/\pi) \times 73.74 \text{ ns}$

$\text{Q1} = 31.22$

T_cutoff_5% $= 3\ q \quad = 3 \times (Q1/\pi\ f0) = 3\ (Q1/\pi) \times T0$

$1500\ ns = 3\ (Q1/\pi) \times 73.74\ ns$

$Q1 = 21.29$

3.3.3. *Measuring Q in the application*

Let us take an example at 13.56 MHz with a quality factor Q1 equal to 31.4, so $Q1/\pi = 10$ and for $3(Q1/\pi) = 30$

Carrier frequency $f0 = 13.56$ MHz

$T0 = 1/f0 = 73.75$ ns

with $Q1 = 31.4$

– consider $(Q1/\pi) = 10$:

T_cutoff_36% $= 0.737\ \mu s$ approx. 10 carrier alternances

– consider $3(Q1/\pi) = 30$:

T_cutoff_5% $= 2.2\ \mu s$ approx. 30 carrier alternances

It should be noted that the value of T_cutoff_5% at 13.56 MHz in this example conforms to ISO 14443 Part 2 A and to NFC ISO 18092 (Figure 3.35).

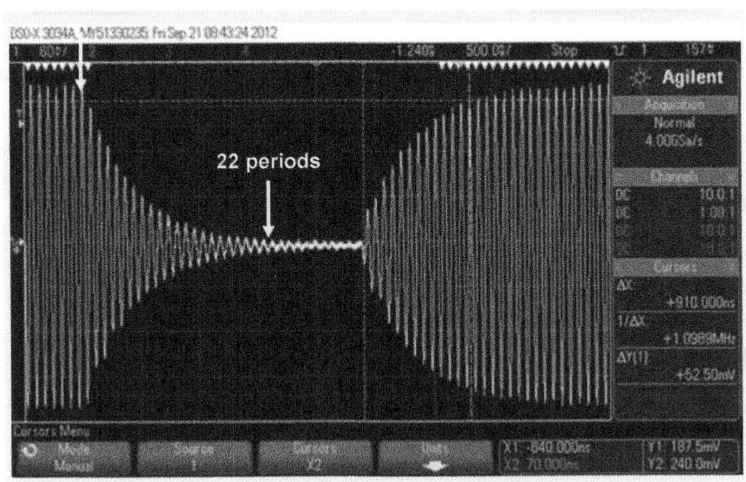

Figure 3.35. *Value of T_cutoff_5% at 13.56 MHz*

PRACTICAL CONSIDERATIONS.–

– Whatever we say or do, the most interesting and only true value of Q is that which is measured in the real application, including the loading effects in the environment (the box, the battery, etc.) which are often difficult to pre-estimate.

– In a mobile environment where a screen and/or a ferrite sheet are needed, a quality factor higher than 35 is very unlikely. When a factor Q of 35 cannot be satisfied in a given environment, it is nonetheless advisable to try and keep it at least equal to 15 in order to obtain decent performances.

3.3.4. *Measurement of the bandwidth in the application*

Beyond the simple equation Bp = f0/Q, which gives an initial indication, the simplest and relatively reliable method is to measure the bandwidth of the antenna circuit of the initiator, with no load, in the real application *via* the frequency spectrum of the amplitude of the radiated field in the air (the "*peak hold function*" of a vector network analyzer – (measures both amplitude and phase properties) (VNA). For this purpose, we can use a small loop antenna of approximately 20 mm in diameter, placed quite a distance away from the antenna so as not to detune it, and just to convince ourselves to measure it also in the presence of a target, to gain an initial idea of the loading effect (Figures 3.36(a) and (b)) – see the Innovision document).

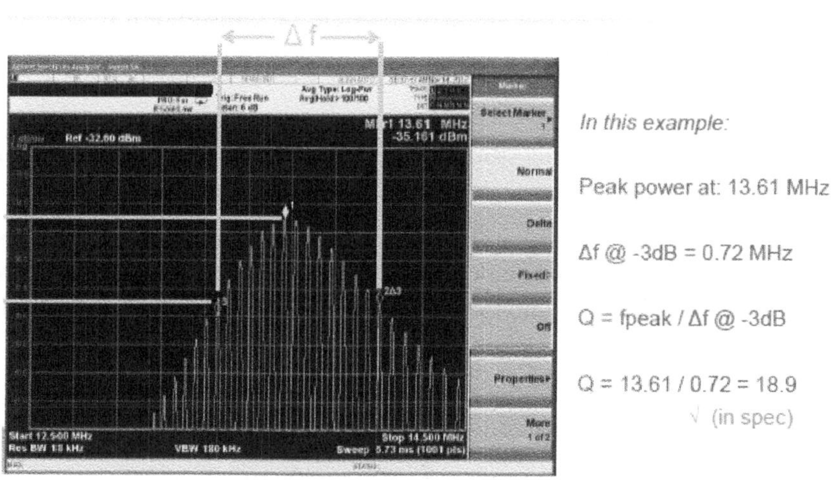

Figure 3.36a) *Measurements of the quality coefficient Q*

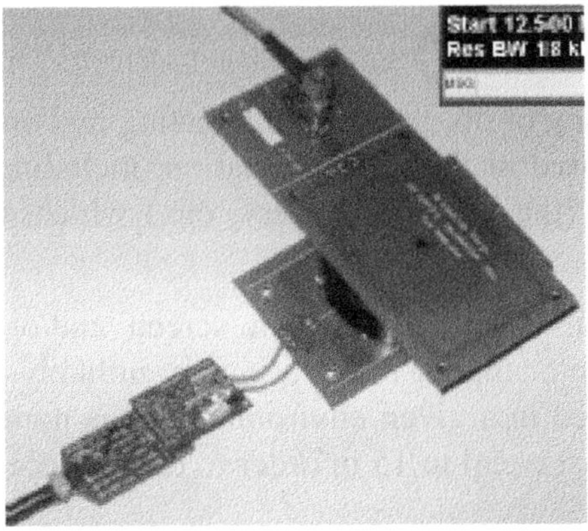

Figure 3.36b) *Measurements of the quality coefficient Q (cont.)*

3.4. Brief handbook on the process of designing an antenna initiator

To summarize, here is a list of the steps – specifically in the correct order – to undertake in dealing with the problem of design and calculation for an initiator antenna … constraints due to the environment apart (screen, loading effect, etc.).

1) Precisely define what target we wish to use in the application (card, mobile phone, USB key, etc.).

2) Know (or determine) the value of the field H_threshold of the target.

3) Define the distance at which we wish to operate (in the air) the target of the initiator.

4) Define the geometric format of the initiator antenna (circular, rectangular, etc.).

5) Depending on the above format, using the Biot–Savart law, as a function of the H_threshold, calculate the value of the field H_0 at zero (two-dimensional (2D) presentation) that the initiator must deliver (without a loading effect or screen).

6) Have a position and know the possible mechanical dimensions of the initiator antenna (and thus its equivalent radius r).

7) Choose/define its number of turns N (even, if possible, to then make it a midpoint) so that the initiator's value Ls is achievable.

8) Calculate the ratio "N/r", and in light of H_0, calculate the value of the current I_Ls circulating in the initiator antenna.

9) In view of the final environment in which the setup will be used, choose or estimate the quality coefficient Q (in the range of max/min values given by ISO 14443 or NFC ISO 18092).

10) For the chosen value of Q, calculate the corresponding value of R_serial.

11) Knowing I_Ls and R_serial, calculate the power dissipated in R_serial which the integrated circuit of the initiator must be capable of providing to the antenna's output circuit.

12) In view of the internal resistance R_ic of the integrated circuit, its supply voltage and configuration (differential/single-ended), verify that it is indeed capable of delivering that level of power.

13) Or else, in another version, for a given supply voltage, calculate the value of R_ic needed to obtain that power.

14) Define/choose the value of freq_cutoff desired to satisfy the ETSI or FCC templates and masks for EMC.

15) Calculate and choose the values of the couple L and C yielding that freq_cutoff, in order to create realistic components.

16) To conclude, calculate the values of the two capacities of the matching T-circuit filter between the filter's output impedance and the initiator antenna's impedance ... and again in order to create realistic components.

4

Examples of Applications of Initiator Antennas

After a long chapter consecrated to the technical and theoretical part concerning the initiator antennas, here is more applications in which, through numerous examples, we will describe the practical concepts of concrete antenna realizations in a very detailed way.

4.1. Large antennas

Why start with large antennas? First, we have chosen the example of large size initiator antennas because they are generally the most complex to carry out (for example, mechanical formats of 10 × 10 to 20 × 20 cm), for example for communicating in high frequency (HF) to 13.56 MHz by inductive coupling between initiators and "NFC device" targets (smart cards, mobile phones, tags, etc.) as well as their command and environment circuits whichever they may be. After all these sections, small antennas will seem ridiculously simple to you.

In the following sections, we will consider two cases of examples of large size antenna applications in communications:

– either with a mono-NFC device in "card emulation – battery-assisted" mode;

– or with multiple NFC devices in "tag batteryless" mode.

4.1.1. *Communication with a mono-NFC device in "card emulation – battery-assisted" mode*

– The "battery-assisted" mode is currently the typical example of communication occurring between an "NFC initiator" and mobile telephones or tablets in the market commerce.

– This supposes that the target functions in battery-assisted mode (normally in this mode, there are none or a few which function by an internal regulator shunt in the NFC device in card emulation mode).

– In card emulation mode, the NFC device target (for example, the mobile telephone) generally possess an offset frequency with quite weak tuning (for example, 14.5 MHz instead of 13.56 MHz) because the device must/is able to pass from the reader mode to the card emulation mode during its use.

– A non-negligible "loading effect" must be seen according to the model of the mobile telephone, its position and distance in relation to the initiator.

– The return communication (target toward initiator) is to this day 95% assured by passive retro-modulation (passive load modulation). Please note that this will change in the near future due to the introduction of the circuits that support *active load modulation* (ALM).

4.1.2. *Communication multi-NFC devices_in "tag batteryless" mode*

Conventional applications are applications of display stand of small articles like jewels in which:

– reading of multiple NFC device targets (of course with management of collisions) of simple target/tags;

– remote power supply of the targets/tags consuming little energy;

– little loading effect;

– return communication (target toward initiator) always assured by a passive retro-modulation (passive load modulation);

– the NFC device target/tag device always functioning in reading mode during its use can possess an offset frequency on its agreement that is absolutely variable according to its application destination (for example, 13.56 MHz up to 20 and some MHz).

4.2. Large antenna in mono-device

4.2.1. *Mechanical formats of the NFC device targets*

These applications are principally those of validator-type initiators of public transport (metro, bus, etc.), chargers without contact with NFC communication, *Point of Sales* (POS) and *mobile Point of Sales* (mPOS), etc., before communicating with mobile telephone or tablet targets, functioning in "emulation card mode – *battery-assisted*" and to this day, whose sizes are mainly defined for the screen sizes equipping them, for example:

– Samsung:

- S5: 14.5 × 7.5 × 0.9 cm

- S6: 14.3 × 70.5 × 0.68 cm

– Apple:

- IP6, 7: 15.8 × 7.8 × 0.9 cm

– Sony:

- Z3: 14.6 × 72 × 0.8 cm

- Z4: 13.9 × 71 × 079 cm

The others follow ... and the overall tendency is in large format.... To when the mobile phone with a screen of 8 inches? (For economic reasons, the mobile telephones "*low cost*" generally have a smaller screen format and therefore a weaker physical size).

4.2.2. *"Form factors" and sizes of antennas of the targets*

Seen from the outer part of an NFC device target, we can neither know the size (very often not respecting ISO antenna classes) nor the real position in which the target antenna is placed because there are no rules between the external size of the NFC target and the physical size of its internal antennas. All the same, to this day (2015) a state of the places of the mechanical size of their antennas can be established.

Before, the antennas were:

– a large size, about 8 × 4 cm to facilitate the reception of a maximum magnetic flux;

– situated somewhere in the rear hull;

– located on a large part of the rear hull of a mobile telephone;

– located on supports in the air and very visible.

Then, there was:

– the antenna implemented on or in the battery;

– therefore with a ferrite foil to magnetically insulate the battery antenna;

– therefore of smaller sizes;

– and even sometimes the antenna is implemented in the SIM or microSIM cards.

The current and future tendency: everybody is trying to reduce the target antenna size (toward SD card sizes, etc.) because to this day, for reasons of quality, the antennas are often etched in technology that is to say in the so-called subtractive technology (loss of Cu or Al) and also loss of support surface (film), which produces a very expensive antenna. Furthermore, with the size of this surface, the shielding/screen of the antenna, via a thin foil of ferrite becomes, too expensive. For all these economic reasons, the tendency is, therefore, having a smaller size antenna ... with a cheaper price, a booster stage in integrated circuit for controlling it, if necessary.

The day after tomorrow the conclusion of the end is not to put an antenna physically anymore and, if it is metallic and waterproof, to serve the metallic casing in the place of the antenna while functioning by Eddy current (as for the RFID luxury Swiss watches in the waterproof stainless steel case). It is a little like what Apple started to do in its iPhone 6 model.

4.2.3. *Application distances required for operation*

In these applications, the physical reading/operating distance between initiator and target can be estimated in several centimeters (maximum 2–5 cm), that is an effective distance between initiator and target antennas of 3–7 cm.

4.2.4. *Estimation of the "loading effects" of the distance or working range*

Before starting to firmly quantify the H magnetic field necessary for good operation, an account of the main causes of "loading effects" present in these applications must be carried out ... because there are always some (we refer you to Chapter 8 for more details).

In the case of target device NFC applications of the mobile phone type *battery-assisted*, although existing, all these loading effects should not be taken into consideration from the target side because the battery location takes into account and manages the consumption differences independently of the power supplied by the initiator field. However, the effects of loading effects should be taken into consideration seriously as regards the initiator side. The visible main effects are large variations of the Q coefficient in loaded conditions and large reductions of the initiator current and coupling index "n" of the system on 13.56 MHz.

4.2.5. *Environment (copper, ferrite, battery, etc.)*

For all the details concerning this realm, we refer you to Chapter 6.

4.2.6. *Several measures for illustrating our proposal*

In order to illustrate our remarks, we take the example of a two turns coil initiator antenna, which is square in shape, about "10 × 10 cm" and whose characteristics are given in Table 4.1.

L	µH	1.22
if Q		25
R = Lω/Q	Ω	4.25
V_batt/sym/50 Ω	V_dc	3.3
P_rms = R I_rms²	mW	45
I_rms	mA	103
N		2
H_rms = N I_rms/2r	A/m	1.82

Table 4.1. *System characteristics*

This initiator antenna mounted asymmetrically is equipped with its complete adjustment circuit (adjustment capacitive bridge and electromagnetic compatibility (EMC) filter) for an adjustment of 50 Ω. This asymmetrical mounting is supplied by a generator of 50 Ω signals between the hot point and the mass of the latter. Moreover, to carry out the measurements of the H magnetic field an antenna of measurement type "*reference calibration coil PICC*" of ISO 10 373-6 (for details, see Chapter 7) is used giving a voltage of V_pp_ref_coil = 900 mV_pp for H = 1 A/m rms averaged over the totality of its surface.

First step: Measurements of field H are, therefore, effected at the initiator antenna level without charge (empty, without target presence – mobile phone – in the field).

Second step: We restart by putting in the field of different targets A, B, C and D (commercial mobile telephones) at the initiator antenna level and the values of the loading effects are noted.

An example of measured values is summarized in Table 4.2.

		Units					
	V_gene_50 Ω = cte	V_pp	10				
	Voltage really applied to the overall aggregate circuit	V_pp	4.1 (… this is due to the fact that the impedance of the entire antenna circuit is not exactly 50 Ω but ~41 Ω!) (practically independent of the loading effects) Information remark: An integrated circuit supplied with 3.3 V_dc and mounted symmetrically would facilitate the optimal adjustment a V_pp voltage really applied in the circuit (3.3 × 4/π × 2) / 2 = 4,2 V_pp or also 1.49 V_rms that is a power of P = 1,49²/40 = 55 mW				
Measurements			Without charge	With charges			
				A	B	C	D
to d = 0	V_pp_ref_coil	V_pp	2.2	0.6	0.7	0.5	0.55
	H_0_rms	A/m	2.45	0.63	0.77	0.55	0.6
	H_0_loaded/H_0_vide	%	–	25	31.4	22.4	24
	V_Rq to the terminals of Rq	V					
	I_ant = V_Rq / Rq	mA					
	Distance "d" to have H_d = 1 A/m	mm	35				
to d = 35 mm	V_pp_ref_coil	V_pp	0.9		0.4	0.2	
	H_d_rms to d = 35 mm	A/m	1		0.44	0.22	
	H_d_loaded/H_d_empty	%	–		44	22	
	Q						

Table 4.2. *Example of measured values*

REMARK.– After having also effected several measurements with a *Voltage Network Analyzer* (VNA) in "Smith abacus" mode, between free space and loading conditions charging tuning, we note that the overall plan of the tuning turns toward the inductive side in the presence of loads.

4.2.7. *H_d field necessary for the NFC device target*

Officially, the NFC ISO 18 092 and 21 481 standards require the initiator to create a minimum field of 1.5 A/m (free space) due to which any NFC device target is supposed to function. That is the theory.

The reality *imposing a constant and uniform H field for the measurement* (in a solenoid mode in order to free itself from loading effects):

– a tag or microchip "light" (style NFC Forum tag T1T or Mifare Ultra Light T2T) often has a sensitivity/threshold close to 0.8/1 A/m;

– a classic card (style DESFire, NFC Forum T4TA) rather 1.3 A/m or for bank applications (Smart XA) 1.4 A/m;

– a Chinese card of unknown technology?

– an NFC operating mobile phone in the "card emulation" mode:

- *batteryless*: it is a great mystery because it depends on what needs remote power supply in the mobile phone so that the application functions (screen, crypto, etc.);

- *battery-assisted*: 0.8–1.1 A/m in order to exceed the operating voltage thresholds of the entry circuits, of demodulation and decoding bits.

Furthermore, to this day the technology of practically all the initiators also operating in voltage/power generator mode than of current generator and the case, at the level of the initiator the battery of the mobile telephones provokes a *loading effect producing a local reduction of the field of 30%, indeed more than in the simple example below (44%)*. That is to say that to succeed in having in "emulation card – *battery-assisted*" mode, the fields of 0.8–1.1 A/m being really necessary when the mobile is present, *one must at least have an H field of 1.1–1.4 A/m free space* otherwise there is a large possibility of it not working. Therefore, to do well, *when starting from the worst standard that is 1.5 A/m, we must dream of creating, with the least words, at the minimum, at the center of the initiator antenna, a H_0 field of 1.8–2 A/m* free space.

4.2.8. *H_0 necessary to create at the antenna level of the initiator*

Here is the first real question of the resolved history:

> "What is the value of the magnetic field H_0 – free space, therefore without load – in A/m provided by the initiator in the center of the antenna?"

This first question and answer immediately lead to the second question:

> "What is the value of the power that the attack circuit can supply?"

Then, the knowledge of the H_0 field will lead to – given for a sensitive target (in V_threshold) and its own loading effect – an operating distance value, then the knowledge of the H_0 field necessary will facilitate the calculation of the H_10 value of the field radiating to 10 m in dBμA/m to verify whether the selected solution is or is not in agreement with the regulation.

4.2.9. *Power P (in watts)*

We start by the power.

During the adjustment of the impedances at stake in the application, there is a direct relationship between the power – in Watt – that can deliver/transfer an integrated circuit and the magnetic field produced in A/m (see a little farther).

In the particular case of the equipment known as "*battery-assisted*" (this is the case here), the receptor part of the NFC target does not (officially) need energy via the field to be supplied and only the value of the flux – therefore, from the magnetic field produced – is necessary to produce a voltage – dφ/dt is, therefore, sufficient for the operating electric threshold of the receptor integrated circuit to be attained.

4.2.10. *Field H which must be produced by the initiator for a specific target*

Besides any coupling and coupling coefficient and any loading effect, we know that, for a rayon r circular antenna (reading antenna), there is a relationship between H_0 (H in the center of the reader antenna) and H_d (H at distance d) which is given one more time by the rationalized theoretical equation of the magnetic field H(a, r) produced by an initiator is in the following form (Biot and Savart law):

$$H(a, r) = \frac{1}{[(1 + a^2)^{\wedge}3/2]} \times H(0, r)$$

where:

– "d" is the distance between the centers of antennas and initiator/tag;

– "r" is the radius of the antenna of the initiator;

– and d = (a × r).

Refer to Chapter 2 for the variations of the relationship H(a, r)/H(0, r) according to the values of "r" and "d", and of the relationship a = d/r for a value of "r".

4.2.11. *Definition of the initiator antenna: format of the "landing area" of the reader (where one puts the target)*

In the light of the mechanical formats of the usual NFC device targets of commerce, in our case, the landing area of the initiator could have a minimum rectangular format of 10 × 10 cm. Furthermore, it is not reasonable to have specific positions by types of mobiles and to indicate them on the surface of the initiator … like the types of airplanes on the tarmac.

4.2.12. *"System" considerations of the application*

The bandwidth (in MHz) that must have the system is a fundamental parameter of the application. As usual, this is linked to the bit rate values, coding bits, modulation types, pause time, etc. The latter are defined and specified in the NFC ISO 18 092 and ISO 21 481 standards. The associated maximum value of the Q quality coefficient of the tuning circuit of the antenna can be obtained from these values.

4.2.12.1. *Quality Q coefficient*

The ISO 18 092 standard evoked above leads to a Q maximum theoretical value of about 35. This value is habitually attained without (too many) major difficulties with antennas free space not subjected to important loading effects. Generally, for many reasons (tolerances, dispersions, operating temperature ranges, drifts in all kinds, aging of components, etc.) and, particularly to facilitate more industrial reproducibility, the producers finally use in the application around 25/30 and, in case the antenna requires a ferrite and supporting screen of the non-negligible loading effects, it is not rare to use to 15/22 ... amortization and physics compel it. The application envisaged is located in this last case:

– with ferrite screen to free it from the environment (and/or without metallic shielding?);

– loading effects due to the various mobile phones.

... therefore, with a Q probable of 15–22.

4.2.13. *Market integrated circuits for direct attack of the antenna*

4.2.13.1. *Existing R_ic_out resistance values*

Generally, the producers of integrated circuits indicated in their data sheets, or their application notes, the value of exit resistance of *asymmetrical* mounting "matching" of which, of course, the value must be multiplied by 2 to have it in symmetrical mounting.

Examples	Producer	Unit	Asymmetrical	Symmetrical
	Inside Secure	Ω	12.5	25
	NXP	Ω	17.5	35
	etc.	Ω	...	...

Table 4.3. *"Matching" exit resistance values*

Using these examples, during impedance adjustments between generator and load, the available maximum power is, therefore, for example, as seen below Table 4.4.

V_batt = 3.3 V	Asymmetrical mounting		Symmetrical mounting	
	R_ic_out_asym Ω	P_eff mW	R_ic_out_sym Ω	P_eff mW
	12.5	44	25	88
	17.5	34	35	68
	25	25	50	50

Table 4.4. *Available maximum power*

4.2.13.2. *Necessary R_ic_out resistance values*

We have shown that all being identical in addition that H = cte √ P

$$H\_eff = \sqrt{(Lo\ N^p)}\ \frac{\sqrt{(P1\_eff\ Q/\omega)}}{N1\ \mu\ s}$$

EXAMPLE.– If the P power increases by a factor of 10, the H magnetic field emitted will increase by a factor of √10 = 3,16 times … that is to say that all things being equal, the classic mounting of reference and EMVCo tests supplied by P = 600 mW (on an equivalent load of 50 Ω) will supply a H_0 field of a value of 3.16 times greater than that supplied with 60 mW by an unfortunate CLCR 663 of NXP of R_ic_out of 35 Ω in symmetrical mounting.

If we know the relationship between H and d and the fact that loading effect “is killing” the magnetic field H_0 of about 30–50%, we can estimate the operating distance “d” in the application.

4.2.13.3. *Conclusions*

At the beginning of this chapter, we have shown that for a symmetrical mounting of impedance adjustment:

$$P\_sym\_eff = V\_batt^2/(\pi^2\ R\_int\_asym)$$

At the given supply voltage V_batt, this means that if we desire to use more power the internal resistance value of the R_ic_out = R_int_asym generator must be reduced.

EXAMPLE.– If we suppose that we want to supply a power of 600 mW (like that required in the mounting of the EMVCo tests):

$$\rightarrow \qquad 0.6 = (3.3)^2/\pi^2\ R\_int\_asym$$

hence $\quad R\_int\_asym = 0.543\ \Omega$

or $R\_int\_sym = 1.09\ \Omega$ (in differential symmetrical mode)

which is the booster circuit case proposed by AMS that we will now examine.

4.2.14. *Booster amplifiers*

An amplifier circuit "booster" (*to boost* means to increase) can be carried out, that is in discrete components, that is in integrated circuits.

4.2.14.1. *Booster amplifier in discrete components*

Figure 4.1 indicates an example of booster stage in discrete components proposed by NXP which we can put at the antenna signal exit out of initiator at 13.56 MHz. This one works in push-pull class A/B (like on audio). This mounting is supplied with *12 V* and consumes about 200 mA when it is active:

– C1 to C4 cut the ongoing;

– D1 to D4 polarize all the transistors;

– RE1 to RE4 assure the middle point of the assembly;

– R1 to R16 equalize the values of the transmitter currents, their tolerances and excesses in temperatures and simultaneously adjust the exit impedance of each way about 25 ohms.

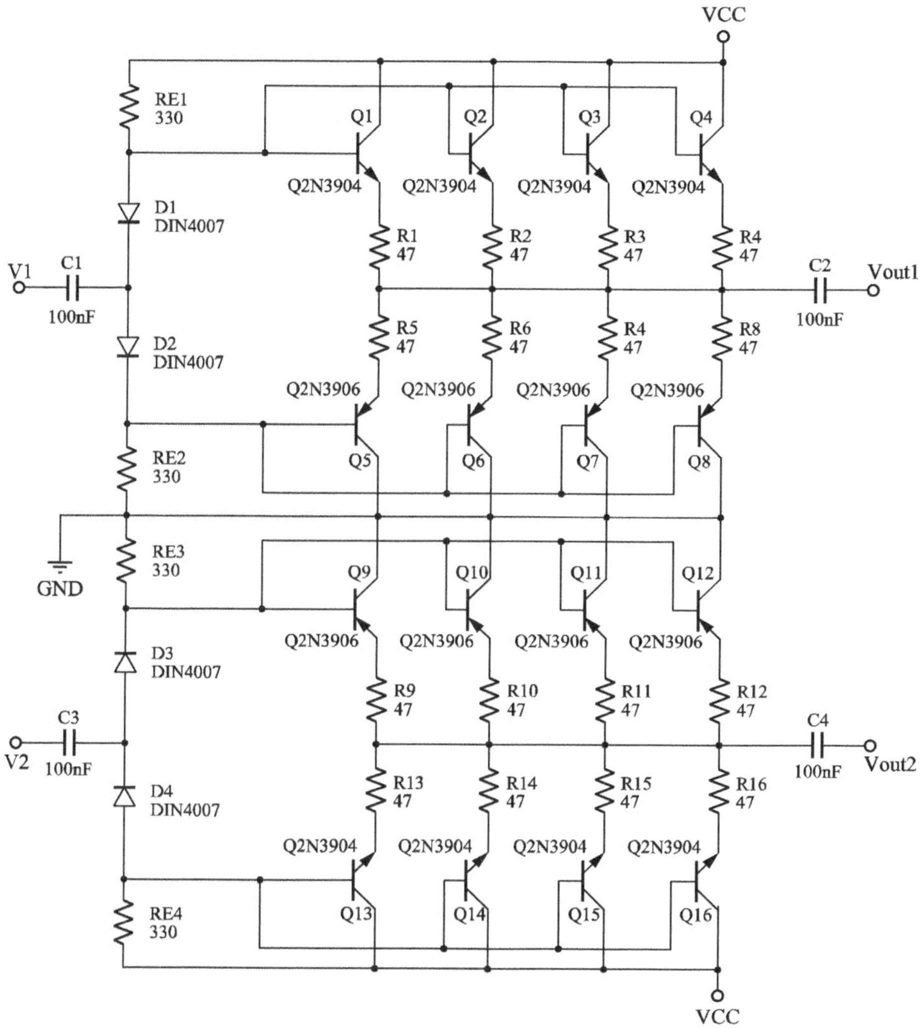

Figure 4.1. *Example of a booster stage in discrete components (source NXP)*

4.2.14.1.1. Conclusion

Despite a favorable technical appearance, this possibility is rarely used because of its number of components, supply value, congestion and additional cost.

4.2.14.2. *Booster amplifier in class C*

Basically, this booster is an amplifier in class C in discrete components, put for example at the exit of a type MiFare Reader Chip (MF RC) xxx circuit.

This architecture (conceived with the Meusonic Company Figure 4.2) requires an HF reception separated by the transmission part (system called bistatic, transmit and reception separated). We must solve synchronization problems between the precision of the value of the tuning frequency of the L circuit, C of initiator power (whose frequency is not linked to the quartz) and that of the demodulation part which, in the integrated circuit, is linked to the quartz for the modulation and the electronic of the coding bit.

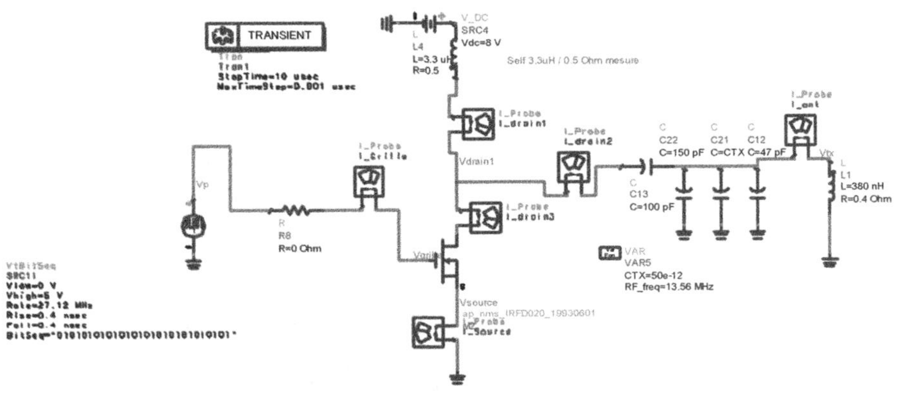

Figure 4.2. *Booster amplifier in class C (source Meusonic). For a color version of the figure, see www.iste.co.uk/paret/antenna.zip*

4.2.14.3. *Booster amplifier in integrated circuits*

Among others, the AMS Austrian company proposes two totally integrated booster circuits facilitating the use of higher power (1 W) in certain conditions. This involves AS 3911 circuits destined to the EMVCo and AS 39230 reader applications to this day in certain mobile telephones.

4.2.14.3.1. AM 3911 and AS 3911B

The AS3911 can function in symmetrical mode with a low output impedance of 1 Ω. With regard to NFC, its main characteristics are:

– ISO 18092 (NFCIP-1) initiator and target (only in active mode);

– ISO 14443 A and B and Felica™ in reader mode;

– supply of 2.4–5.5 V (3.3 and 5 V modes).

Available power:

– 1 W maximum in differential mode (symmetrical exit);

– 200 mW maximum in single-ended output mode (asymmetrical exit);

– automatic selection of ALM and Passive Load Modulation (PLM) (I/Q) demodulators.

The maximum power of 1 W facilitates the satisfaction of certain applications joining antennas of (very) small size of card types microSD, SIM cards, small dongles in which the environment is hostile and the spaces are limited.

Figure 4.3 gives an example drawn from an AN of at AMS for applications of EMVCo-type reader whose antenna is the two turns Europay Mastercard Visa (EMV) classic antenna, diameter 7 cm and of L = 590 nH.

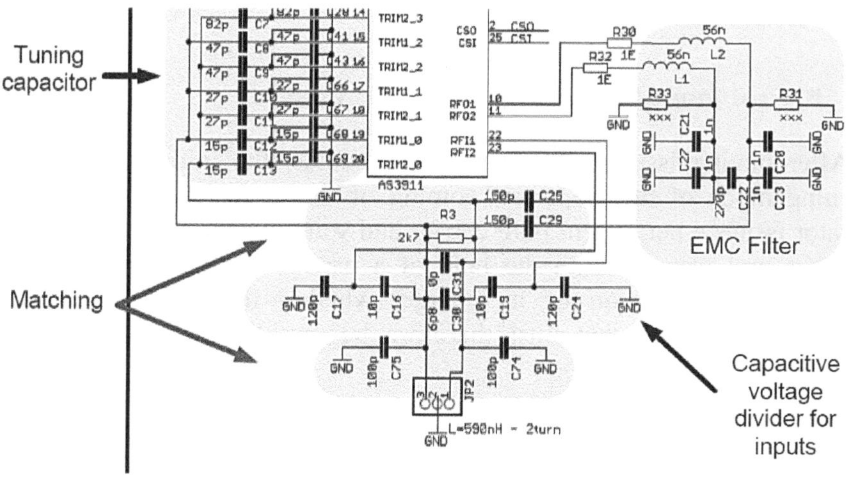

Figure 4.3. *Booster amplifier in integrated circuits (source AMS)*

4.2.14.3.2. AM 39230

To our knowledge, the AM 39230 circuit is currently only available to telephone professionals.

While adding on one hand a booster function in the analog frontal of a classic NFC controller and in implementing on the other hand a modulation of active load ALM on average increasing the return signal of retro-modulation of the target toward the initiator of a 20 factor in relation to a classic PLM method (the NFC device target is in mode tag, card or card emulation), this circuit offers more of interoperabilities and performances. These two improvements allow for the antenna concept whose size can be reduced up to about 100 mm^2 (of the μSim), which is therefore less burdensome than of the classic solution of ID1 format. Furthermore, the use of ALM also allows for the use of NFC in environments that are difficult such as metallic cases while envisaging to remain compatible with the great conformity standard of the ISO market, EMV (bank), CEN (transport).

In conclusion, knowing that the consumer market will use its mobile telephone or "*smart watch*" to carry out payment without contact in the shops only if this technology works or will not use tickets in the mass transport infrastructure, access control as well as will not support "*wearables*" (smart watches, bracelets, etc.) and other clothing items. Whatever the manner of presenting the target (in card emulation mode *battery-assisted*) to the initiator is, this solution with ALM presents numerous advantages.

4.2.15. *Problem of the retro-modulation value*

After having issued interrogation commands, the initiator passes into listening mode of the responses coming of the target. To this effect, the initiator issues a permanent pure carrier and waits for the target to signal its presence and respond with the help of a particular retro-modulation that represents the variation of its charge, whether the latter is active or passive, of little level that must, therefore, be amplified (see Chapters 3 and 6 for details in PLM/ALM).

4.2.15.1. *Amplify the signal received/preamplifier stage*

It may be that the return communication cannot be established because, despite the fact that the target receives enough energy from the initiator to be remotely fed or, although it be *battery-assisted*, and the initiator cannot interpret its messages. In this case, it can be necessary to amplify the signal received when putting a preamplifier of which there is an example of a diagram again originating from NXP given in Figure 4.4, between the reception antenna and the Rx entry of the integrated circuit of the initiator.

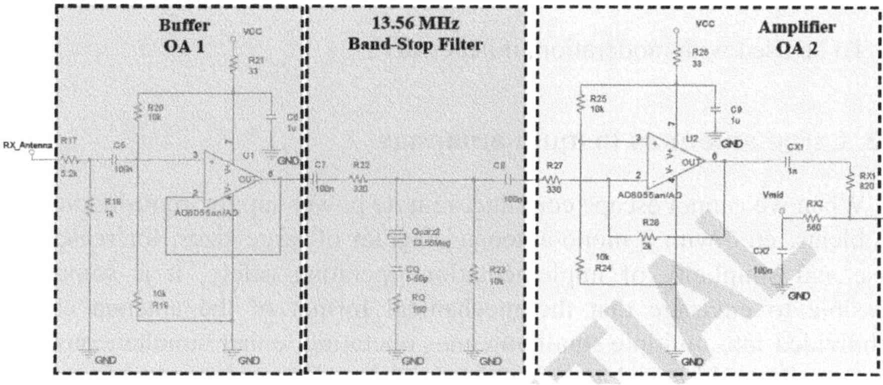

Figure 4.4. *Preamplifier stage (source NXP)*

The goal of this circuitry is to reject the sinusoidal signal of the carrier at 13.56 MHz while only allowing the two bands of the retro-modulation subcarriers located on both sides of the central frequency. The diagram, therefore, is composed of a frequency rejecter filter made of a quartz at 13.56 MHz inserted between the central point of two capacities of liaison and the mass. The quartz, functioning in its series resonance frequency, then becomes a short-circuit and traps, therefore, with very great precision the 13.56 MHz originating from the Rx part of the antenna.

The coefficient/the depth of rejection of the filter are adjusted with the help of a low value series resistance with the quartz. The bandwidth of the filter is adjusted to 1 MHz.

Note that this stage is very close to a Colpitts oscillator and we must protect ourselves from guaranteeing its stability while muzzling the Q of the quartz with a resistance as indicated above.

After the follower first stage of gain 1, when R_series = 0, the quartz filter lowers the 13.56 MHz of –25 dB in relation to the filter entry and the

14 MHz of –2dB, that is a gap of 23 dB between 13 and 14 MHz. The second operational amplifier (op-amp) has an intrinsic gain of 5.46 dB (R2/R1 = 2 k/330 ohms = 6), and at the exit of the stage a resistive divider bridge is located on the Rx pin, which therefore gives an overall gain in voltage of 3.5 in total. Out of the band (filter and amplified), the 14 MHz is, therefore, recovered from 1 dB.

To be used with moderation, if necessary.

4.3. Large antennas in multi-antennas

When we cannot escape coupling, remote power supply, retro-modulation problems, etc., with a mono-antenna diagram of large sizes, for reasons of ease and simplicity of implementation operating safety, it is sometimes possible to envisage that the mechanical format of the antenna can be subdivided into multiple small antennas operating, either simultaneously, or temporarily multiplexed and of which it is also always necessary to envisage the economic aspect of the solution:

– 2 small antennas + 2 small known circuits → not expensive;

– 1 large antenna + 1 circuit → more expensive + acrobatics.

4.3.1. *In simultaneous mode (temporarily non-multiplexed)*

4.3.1.1. *Division of the mono-antenna of the initiator in “n” antennas*

In order to succeed in covering the totality of the desired surface of the initiator, we examine the solutions consisting of dividing the initial large antenna into a (small) network of 2 or “n” antennas in series and/or parallel mounting, each of a smaller size in order to increase the coupling between reader antenna and tags and to stimulate them all together.

4.3.1.1.1. Flow addition and subtraction problems and zones of “zero lines”

This simultaneous operating mode of several antennas in the same perimeter is not easy to implement. In fact, pursuant to the superposition theorem, on one hand according to the direction of the winding of the coplanar antennas and on the other hand, currents circulating simultaneously in two branches of two close loops one from the other, the magnetic flux generated by the initiator can be added locally using algebra, be subtracted and even be canceled from loop to loop and from coil component to coil component because the magnetic flux emitted by each antenna of the initiator at least induces the adjacent antennas by simple coupling of proximity (of coupling coefficient "k") leading to more or less harmful effects.

Although, in the case of coplanar antennas, the coupling coefficients between antennas have weak values, all the same one must pay attention locally to the connection of the antennas and the flux direction at risk of field cancellation and "zero lines" zone creations in the common branches.

4.3.1.1.2. *Incidences due to voluntary or involuntary coupling of the antennas belonging to the same branch or to neighboring branches*

Figure 4.5 illustrates different operating cases.

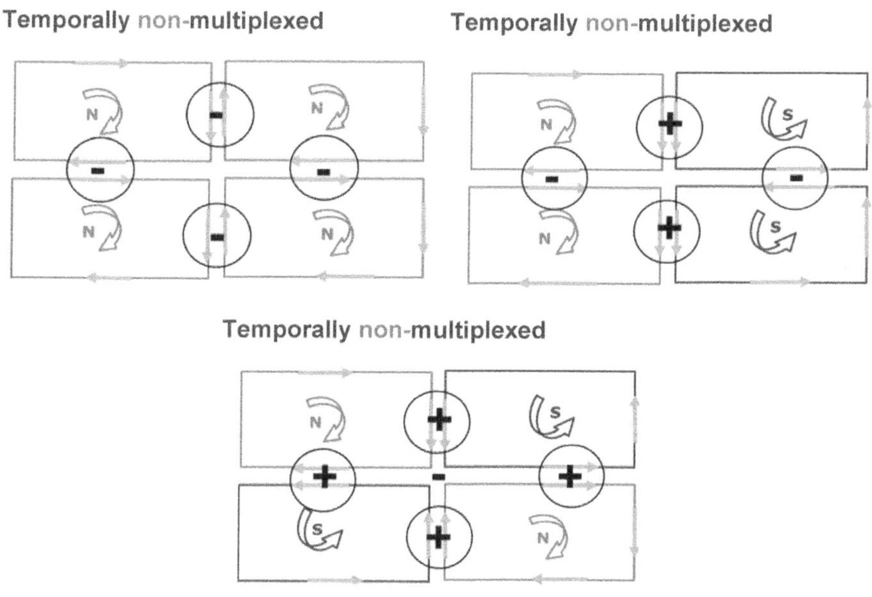

Figure 4.5. *Branch to branch incidences*

4.3.1.2. *Division of the antenna in 2 or "n x n" antennas in parallel-series mounting*

See the details of this principle in Chapter 6.

For example, let us take a network composed of four coplanar, identical antennas, of a mechanical format noticeably a quarter of the total surface and which are sufficiently separated from each other in a way that the magnetic couplings can be considered as negligible (Figure 4.6).

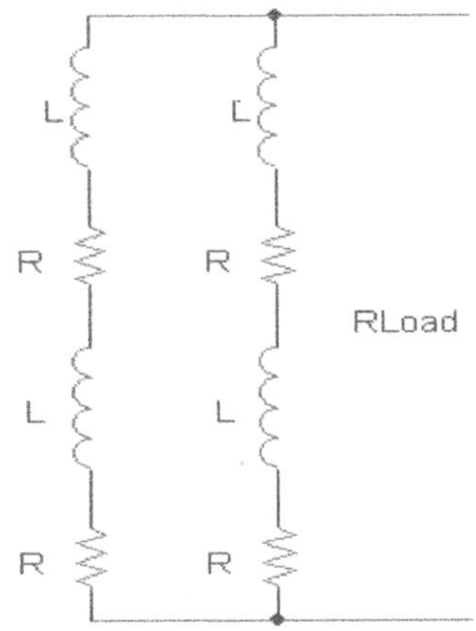

Figure 4.6. *Antenna principle "n x n"*

Each of these four new antennas has electric values (L and R) strictly identical to the value of the antenna of a solution mono-antenna. The trick of mounting consists of the new impedance of this network being strictly equal to the impedance of the initial antenna and that it does not change anything entirely with regard to impedance adjustments. In fact, the value of the impedance equivalent to this antenna network is L in the beginning (putting in parallel two branches composed of two Ls connected in series). Note that this principle can be evidently generalized to parallel "n branches" each composed of "n identical antennas", which are/themselves arranged in a series.

4.3.1.3. *Conclusions*

Despite a favorable technical appearance, this possibility falls short because on the whole:

– better geographical division of the fields;

– the initiator current is divided by two (two parallel branches);

– in the same L, the flux Φ = L I generated by antenna is, therefore, half;

– at the same L value, the antennas are dimensionally smaller and we can try to make up for the loss of surface of each of them by increasing the number of turns of each antenna that favors the N I = H product;

– difficulty of fighting zero lines despite the judicial choice of the direction of the rolling-up and generated flow, leading to certain positions juggling of the tags in the surface to be covered;

– weak additional cost if not null.

4.3.2. *In multiplexed mode temporarily*

In this case, we go back to known problems more or less. This case is easier to manage with regard to the effects of cancellation of branches and fields and we can even dream of doing a little surface covering to fill up field holes. This solution requires circuits of the command HF NFC standard + the multiplexing management generally by bus series type I2C or SPI + the management software of the communication exchange, but all that is not very complicated.

Via several ideas of solutions already proposed and executed on the market, we now detail this multiplexed mode.

4.3.2.1. *Temporarily multiplexed antenna network*

In order to cover the whole surface, we can carry out an antenna network temporarily multiplexed (therefore, one after the other) via a bus of SPI/I2C style with the help of two or several antennas actives (2, 3, 4, etc.) with each having a specific circuit of initiators (for example, CLCR 663).

The solution (Figure 4.7) with only two antennas can be very well adapted and adopted for an initiator which must cover, for example, a surface of 16 × 8 cm which is the example type of this kind of simple temporary multiplexing by pure and simple juxtaposition of 2 times the same concept on the copper.

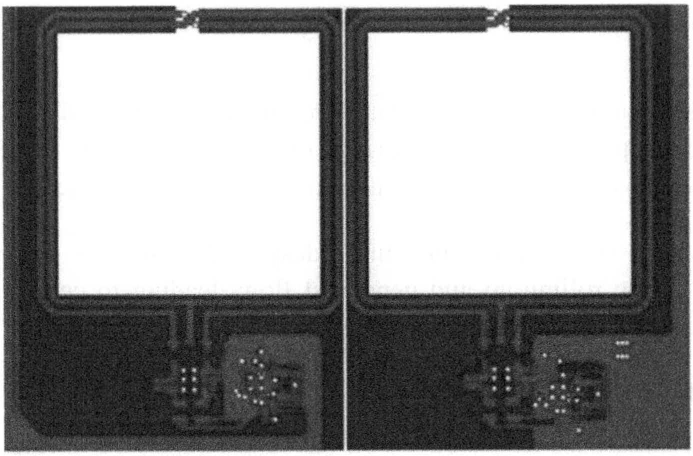

Figure 4.7. *Juxtaposition of 2 times the same concept on the same copper*

4.3.2.2. *Temporarily multiplexed antenna network with cover*

A partial overlay of antennas (Figure 4.8) can also be envisaged to cover the central area more precisely. In this case, we must think of "retuning" the two antennas because of the interaction of the antenna which is active while the other one is not supplied.

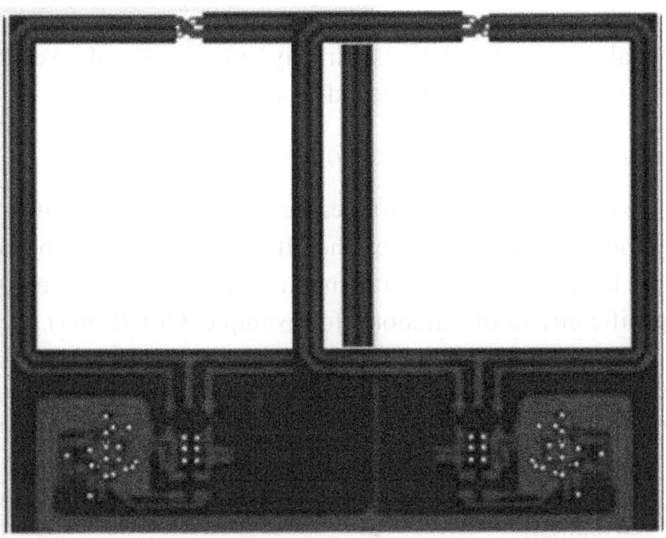

Figure 4.8. *Partial cover of antennas*

4.4. Large antennas in multi-devices

The solution (Figure 4.9) permits the covering of a surface of 20 × 20 cm by a temporal simple multiplexing and a pure juxtaposition of 4 times the same concept on the same copper. This solution (Figure 4.9) can be envisaged for reading applications of presence and identification of different small objects (in jewelry, etc.) collision and retro-modulation management problems of each are easier to resolve.

Figure 4.9. *Juxtaposition of 4 times the same concept on the same copper*

For example, in the case of the use of four temporarily multiplexable antennas, several non-limiting examples/eventualities can be envisaged (Table 4.5).

Number of antennas	Forms and connections	Multiplexing mode	Number of command integrated circuits	Surface cover of the antennas	Winding and flow direction
4	Identical forms and values All taken individually	In four phases each antenna alternately	4	Without	All in phase
					Two by two in opposition of phase
4	Identical forms and values Mounted in series two by two	In two phases per couple/ antenna branches	2	Without	In phase
					In opposition of phase
				With	In phase
					In opposition of phase

Table 4.5. *Temporary use of four multiplexables antennas*

The main characteristics of mounting (see Figure 4.10) used in the first place are:

– the flux direction (here all the winding is in the same direction);

– the cancellation of the fields on the branches adjoining/juxtaposed cannot be produced since the product fields for each antenna are not simultaneous;

– in order to cover the entire surface, the antennas are/must be situated very close to each other, without (*of course*) being disturbed by time frames between active (without “s”) and passives (with “s”);

– when an active antenna functions, it has two close and coplanar antennas on both sides, which are therefore passive but granted. Due to the coupling (k even weak), these passive antennas, therefore, participate in light *detuning (see below)* of the antenna at this active moment. We must, therefore, think of readjusting the tuning of the latter a little with the help of T filter capacities. During calculations (see below), because of coupling between antennas, the value of the inductance of the active antenna tends to diminish a little, which noticeably increases the values of these capacities.

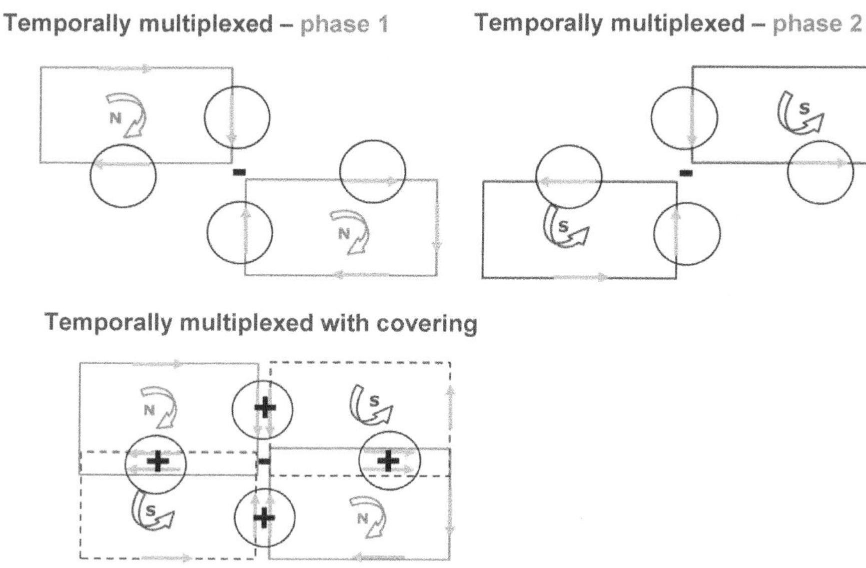

Figure 4.10. *Interactions between antennas*

4.4.1. *Conclusions*

Despite a favorable technical appearance, this possibility falls short because:

– better geographical spreading of the fields;

– the current of the initiator is not divided by two;

– at the same value of L, the flux Φ = L I generated by the antenna is, therefore, identical;

– at the same value of L, the antennas are smaller and we can try to compensate while increasing the number of turns of each antenna favorable to the product N I = H;

– it is easier to fight the zero lines;

– weak additional cost, if not null.

4.5. Other examples of initiator antennas

There are also many other tricky diagrams of antennas. To mention but a few:

– monostatic initiator antenna but duplicated, facilitating successful remote power supply to all the necessary parts to a secured transaction (SE Secure Element, etc.) of an NFC device target of "*battery-assisted*" even when its battery is withdrawn (namly "out"). The possible examples of the block diagrams of the PN 544 and PN 7120 of NXP and their antennas are given in Figures 4.11(a) and (b);

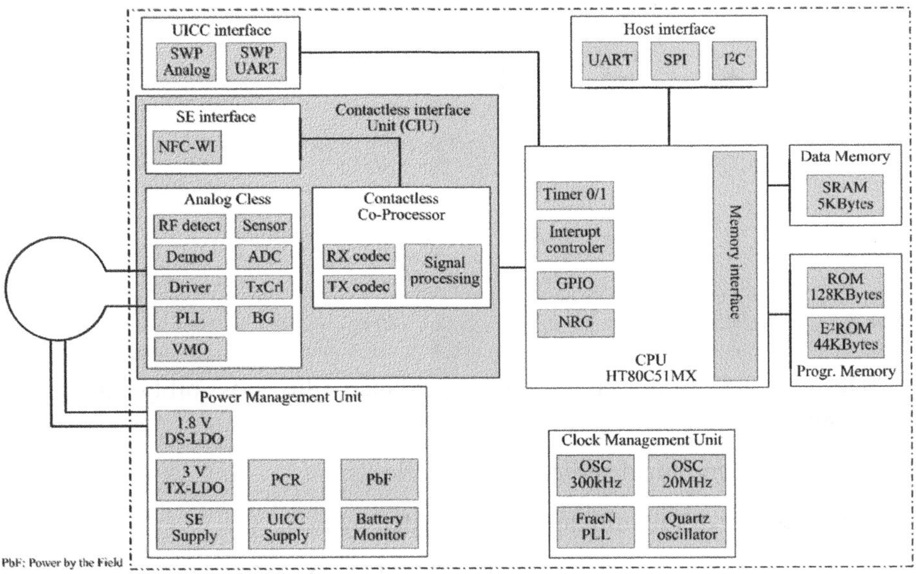

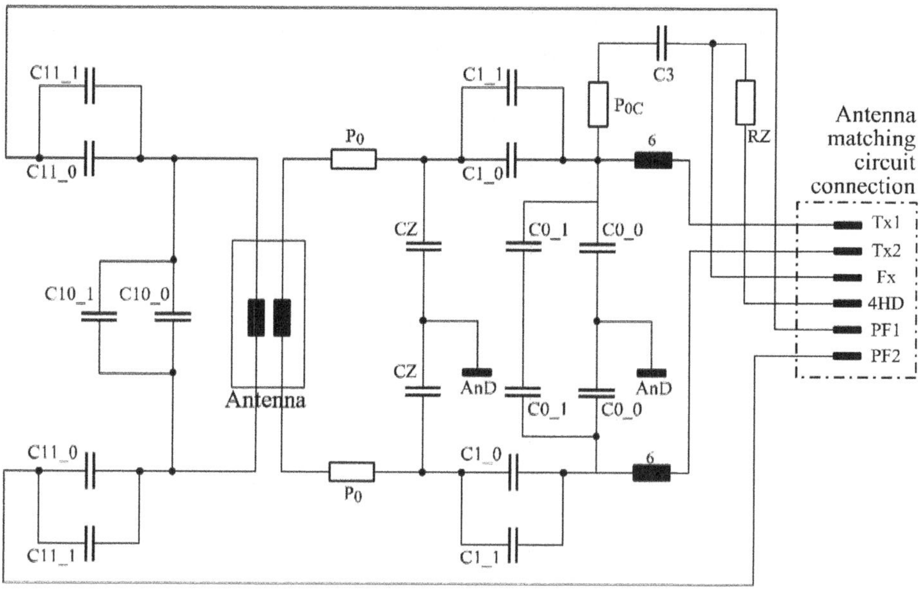

a)

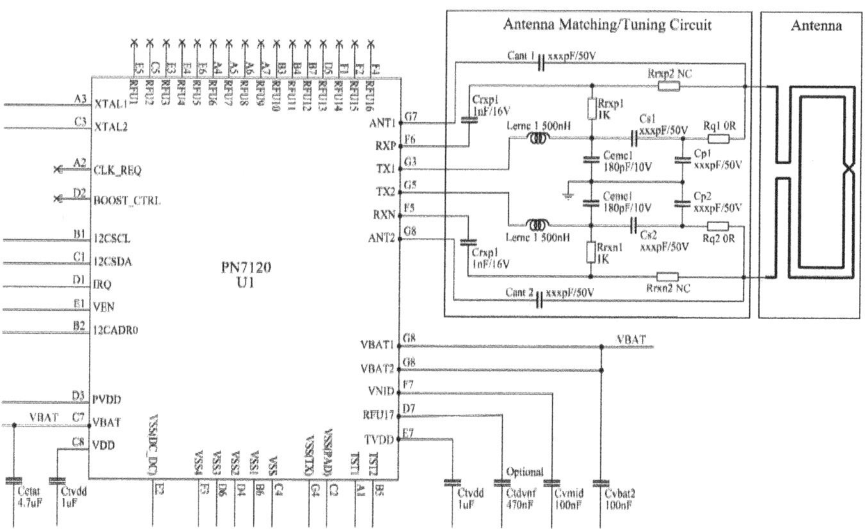

b)

Figure 4.11. *a) Block diagram of the PN 544 (source NXP); b) block diagram of the PN 7120 (source NXP)*

– mounting by an auto-adjustable tuning system of the antenna to compensate the possible lack of tuning due to loading effects of the targets;

– diagrams purely of the "bistatic" type having two distinct antennas to the initiator, one for the transmission, the other for the reception. Despite the fact that they are distinct, the latter can be coupled, etc.;

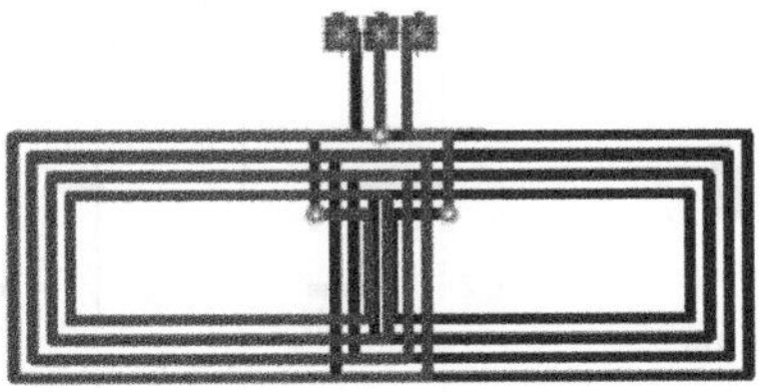

Figure 4.12. *Diagrams of the "bistatique" type*

After these long episodes consecrated to the antenna "initiators", at present we refer you to the antennas "targets".

5

Antennas for Targets and Tags: Detailed Calculations

This third chapter of the second part aims to assist the antenna concept of the "NFC device TARGET". It is divided into three sections:

– antennas for passive targets;

– antennas for tags (NFC ForumTags);

– antennas for targets in card emulation mode.

5.1. Introduction: there is a target and target

Before starting this chapter, we present a detailed inventory of the different NFC device "targets" that underlie the international standards (ISO and CEN), the propriety standards (NFC Forum, EMV, etc.) and other babble propagated by the press.

For this purpose, these "targets/tags" must be finely split according to the three main parameters which are the modes, supply types, retro-modulation principles, which will count, for each case, "targets/tags" antennas of different designs (Table 5.1).

5.2. NFC Forum Tags

Before starting, let us return for a moment to the NFC Forum devices of the NFC Forum which has formalized the distinctions of its "NFC Forum Tags" a little more while creating technology subsets.

<table>
<tr><th></th><th colspan="2">NFC mode of communication of the target</th><th colspan="2">Type of supply of the target</th><th colspan="2">Retro-modulation mode of the target toward the initiator</th><th>Examples of applications of targets</th></tr>
<tr><td>NFC device “Target”</td><td>Passive</td><td>Active</td><td>Battery-less</td><td>Battery-assisted</td><td>Passive</td><td>Active</td><td></td></tr>
<tr><td rowspan="2">Ordinary “Target”

ordinary tag</td><td>x</td><td></td><td>x</td><td></td><td>x</td><td></td><td>Standard label.
Simple smart card “access”, “fidelity”.
Poster NFC Forum, etc.</td></tr>
<tr><td>x</td><td></td><td></td><td>x</td><td>x</td><td></td><td>Badge for museum visit with sound comments, etc.</td></tr>
<tr><td rowspan="2">Ordinary “Target”

Card emulation</td><td>x</td><td></td><td></td><td>x</td><td>x</td><td></td><td>Mobile phone in payment smart card emulation mode</td></tr>
<tr><td>x</td><td></td><td></td><td>x</td><td></td><td>x</td><td>Mobile phone in payment smart card emulation mode</td></tr>
<tr><td>“Target”

Peer-to-peer</td><td></td><td>x</td><td></td><td>x</td><td colspan="2">Active mode transmission in each direction</td><td>Exchange between an initiator and a target in Peer-to-peer mode</td></tr>
</table>

Table 5.1. *... there is a target and target*

5.2.1. *“Technology Subset”*

The NFC Forum divided NFC-A technology, NFC-B, NFC-F already presented from the “Technology Subsets” (subdivisions) in which the “NFC Forum Tags” which are compatible targets with one of the four “Technology Subsets” (and their associated protocols) or else a simple target in accordance with the ISO 18 092 – NFC IP1 standard. Furthermore, for reasons of heritage, the four following “Technology Subsets” were based on products existing on the market (MiFare, FeliCa and Topaz) or purely in conformity with the ISO 14 443 standards.

The following tables summarize and formalize the state-of-the-art to this day (May 2015) in the NFC Forum (Table 5.2).

Table 5.3 details the general properties of each of these tags.

Table 5.4 details the general properties of each of these tags.

<table>
<tr><td></td><td colspan="8">Technologies</td></tr>
<tr><td>Activities</td><td colspan="4">NFC – A
based on
ISO 14 443 – A</td><td>NFC – B
based on
ISO
14 443 – B</td><td colspan="2">NFC – F
based on
JIS X6319-4</td><td>NFC – V
based on
ISO
15 693
In the course of editing</td></tr>
<tr><td>Listen, RF Collision Avoidance, Technology Detection, Collision Resolution</td><td colspan="4">*based on*
ISO 18 092 – NFC IP 1</td><td>*based on*
ISO
21481
NFC IP 2</td><td colspan="2">*based on*
ISO 18 092 – NFC IP 1</td><td>*based on*
ISO
21481
NFC IP 2</td></tr>
<tr><td rowspan="6">Device Activation</td><td rowspan="9">NFC-DEP Protocol
based on
ISO
18,092</td><td colspan="3">NFC – A</td><td>NFC – B</td><td>NFC – F</td><td rowspan="9">NFC-DEP Protocol
based on
ISO 18 092</td><td>NFC – V</td></tr>
<tr><td colspan="5">*Technologies Subsets*</td><td></td></tr>
<tr><td>Type 1 Tag Platform</td><td>Type 2 Tag Platform</td><td>Type 4A Tag Platform</td><td>Type 4B Tag Platform</td><td>Type 3 Tag Platform</td><td>Type 5 Tag Platform</td></tr>
<tr><td colspan="5">*Industrial products and heritage protocol*</td><td></td></tr>
<tr><td>*Topaz*</td><td>*Mifare UL*</td><td>*ISO*
14 443 – A</td><td>*ISO*
14 443 – B</td><td>*FeliCa*</td><td>*ISO*
15 693
(2015)</td></tr>
<tr><td colspan="5">*Communication protocol*</td><td></td></tr>
<tr><td rowspan="2">Data Exchange</td><td colspan="2" rowspan="3">Type 1, 2, and 3 Tag Half-duplex Protocol</td><td colspan="2" rowspan="3">ISO-DEP Protocol
based on
ISO 14 443 (- 4) and EMV_CLESS</td><td rowspan="3">Type 1, 2 and 3 Tag Half-duplex Protocol</td><td rowspan="2"></td></tr>
<tr></tr>
<tr><td>Device Deactivation</td><td></td></tr>
</table>

Table 5.2. *NFC Forum – "Technology Subset"*

NFC Forum Tag	so-called	Techno	Heritage	Size	Debits kbit/s	Example of products
Type 1 Tag	T1T	NFC – A Without management of the collisions	ISO 14 443-A	96o to 2ko	106	*Topaz (of Innovation)* simple tags of read/re-write or read-only type
Type 2 Tag	T2T	NFC – A With management of the collisions	ISO 14 443-A	48o to 2ko	106	*MiFare UltraLight (of NXP)* read and re-write and configurable tags in read-only up to 2 Ko of memory
Type 3 Tag	T3T	NFC – F With management of the collisions	JIS X 63194	< to 1Mo	212 – 424	*FeliCa (of Sony)* read and re-writable preconfigured tags, or read-only limited to 1 MByte per service.
Type 4 Tag	T4AT T4BT	NFC – A NFC – B With management of the collisions	ISO 14 443-A ISO 14 443-B	32ko	106	*compatible ISO 14 443 (part 2, 3 type A & B and part 4) – EMV CL and ISO 7816-4 low layers and Peer to Peer protocol,*
Type 5 Tag	T5T	NFC – V	ISO 15 693 Edition 2015		Qq kbit/s	*the NFC Forum is in the process of validating the "NFC-V" technology and tags so-called "T5T" corresponding to the ISO 15 693 standard of Vicinity Cards*
Type 6 Tag *For information!*	*T1T* *T2T*	*NFC – B* *With management of the collisions*	*ISO 14 443-B*			*the current specifications of the NFC Forum do not describe simple tags of type 1 and 2 in version ISO 14 443 B*

Table 5.3. *NFC Forum – Tags…*

(May 2015)	ISO 14 443 A	JIS X 63194 (FeliCa)	ISO 14 443 A	ISO 15 693
Standard	ECMA 352 NFC IP 2			
Standard	NFC Forum			----
"Technology"	NFC – A	NFC – F	NFC - B	NFC – V
Standard	ECMA 340 NFC IP 1			
Coding bit				

Ascending link	Modified Miller	Manchester	NRZ	In the course of definition
Descending link	Coded Manchester under carrier	Manchester without standard under carrier	NRZ coded BPSK	
Debit kbits/s				
Ascending link	106 Only	212 and 424	106 Only	
Descending link	106 Only	212 and 424	106 Only	
Carrier modulation				
ascending modulation	ASK 100%	ASK 8 à 30% 8 à 14%	ASK 8 à 14%	
descending modulation	OOK Load Modulation	OOK Load Modulation	OOK Load modulation	
	PCD & PICC			
Magnetic field				
Ascending link				
H_min rms	1.5 A/m unavailable	1.5 A/m Unavailable	1.5 A/m unavailable available	
H_max rms	7. 5 A/m Unavailable	7. 5 A/m Unavailable	7. 5 A/m unavailable	
Retro-modulation				
Descending link				
H_min rms	Unavailable	Unavailable	unavailable	
Techno subsets	T1T T2T T4AT	T3T	T4BT	T5T

Table 5.4. *NFC Forum versus ISO*

5.3. Introduction to problems of antenna targets/tags

The *near field communication* targets/tags and contactless equipment (of proximity or vicinity to 13.56 MHz), of all "*form factors*" and all standards, all operate in the "Air Interface" according to a similar principle mainly based on the presence of a resonant tuned circuit between an L inductance and a C capacity permitting us to obtain a minimum voltage (V_threshold) sufficient for activating the communication and functional integrated circuit placed on this equipment.

5.3.1. *Tuning of the targets/tags*

The usual custom for carrying out this tuning consists of using an L, C-type circuit, and of tuning it on the incident frequency called carrier frequency. The tuning frequency of this circuit is subjected to the classic relationship $f = 1/(2\ \pi\ \sqrt{LC})$ and, for example, in the case of tuning to f_carrier at 13.56 MHz, ω = 85,157 10^6 rd/s, the condition $L\ C\ \omega^2 = 1$ is filled when the value of the product LC = 137.8, when the inductance L value is expressed in µH and that of the C capacity in pF. Table 5.5 gives some examples of the possibilities to be carried out for the two values L and C for perfect tuning.

C pF	L µH
15	9.2
30	4.6
60	2.3
100	1.37
150	0.92
220	0.63
330	0.42
390	0.35

Table 5.5. *Value of the coupled inductance L in µH and capacity C in pF*

5.3.2. *The inductance L*

Winding constituting the L inductance of a piece of contactless equipment is usually carried out using a wire, printed, flexible circuits/films, vaporized deposits of copper or aluminum, of deposited conductive inks or else flashed inks. Generally, to a certain extent, we can choose a value of the

L inductance that is most suitable to the situation, to the important details regarding mechanical burden (for example, antennas of small sizes for the ISO 14 443 card formats of classes 4 to 6, that is to say close to the SIM card or µSIM) and that the electromagnetic environment present around the winding (metal, case, battery, etc.) is often fixed. In reality, in these applications, the inductance value is close to the desired value but not absolutely that which was desired.

OBSERVATION.– At 13.56 MHz, two important things must be remembered from Table 5.5:

– the maximum value of the inductance is relatively limited (around 10 µH) because of the reproducible industrial physical minimum capacity (~10 to 15 pF are usually values considered sufficient for the eradication of all the parasite and dispersion capacities);

– the minimum value is also limited (~ 300 nH) due to small sizes of the winding realizables on surfaces such as µSIM cards;

– that is, in summary: ~ 300 nH < L < ~ 10 µH … and we must manage with that.

5.3.2.1. *The capacity C*

In order to tune the L inductance, we must have a C capacity.

5.3.2.1.1. Nominal value

Very often, due to the physical and mathematical relationship (relationship which is one of the pillars of this profession) linking the minimum intensity of the magnetic field H_min (in A/m) to the value of the applied frequency in contactless equipment (labels, tags, etc), the usual habit is to say that the L, C circuit must be tuned on the incident frequency of the carrier, but in reality, generally, it is not absolutely the case for very specific reasons.

In fact, the choice of the value of the L, C tuning frequency, is not indifferent and pertains to strategic application choices. Despite the theory optimum of energy transfer that is tuning the frequency of the target/tag to that of the carrier frequency of the incident wave, it is interesting, preferable, desirable that the tuning of the target possesses a voluntary offset (difference) of the tuning frequency toward the larger frequencies (between 15 and 19 MHz) (see Figure 5.1). This allows us to tolerate certain hostile effects that are always present (hand effects, presence of battery, ferrite screen, etc.) and the presence of other targets arranged in stacks (in piles, cards in a portfolio, numerous visas in a passport, etc.) The value of this offset (small or great) is left to the choice of the developer (as indicated in the example of Figure 5.1) because it is structurally a function of the application and this often implies that, for a similar design of a product and of the L value of its antenna, we are obliged to arrange different capacity values according to the clients and their applications (for example, at 13.6 MHz of some 10–390 pF).

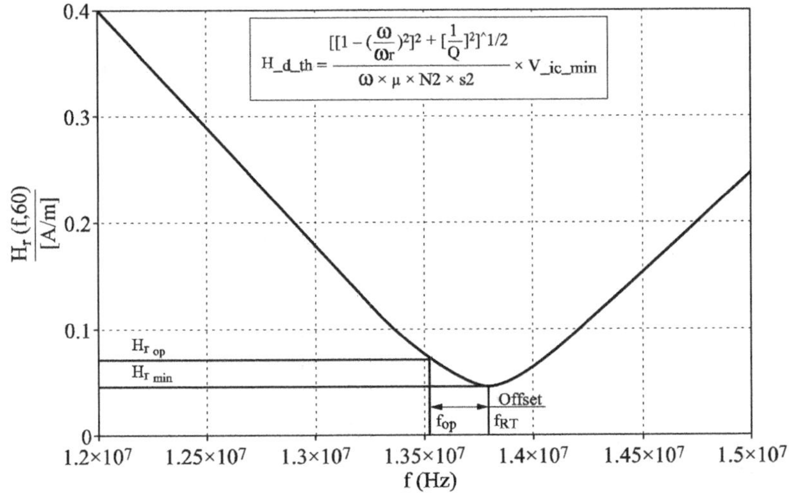

Figure 5.1. *Magnetic field H based on the tuning frequency*

There are also numerous technological possibilities for satisfying these last values that, as usual, individually have their advantages and limits. In fact, these capacities can either be:

– integrated in the silicon of the integrated circuit with (pre)defined values (often of 20, 50, 70, 90 pF according to the types, associated markets and manufacturers), with tolerances, of the electric qualities, etc.;

– totally discrete and external to the integrated circuits not having internal capacities or by a conventional component of 20–400 pF, or by a capacity carried out locally in serigraph technology (of 20–60 pF);

– by juxtaposition of integrated capacities and discrete or serigraphed external additional capacities.

5.3.2.1.2. Precision

Nominal values also mean precisions and tolerances.

The discrete capacities generally have smaller tolerances (+/– some %) than the serigraphed capacities (the drawing produced by photoengraving and the coefficient of the dielectric only change a little, only the thickness of the dielectric will be controlled) and, in conclusion, the integrated capacities of the integrated circuits generally have tolerances of +/– 15%. The tuning frequency of the different contactless components being subjected to the relationship $f = 1/(2\ \pi\ \sqrt{LC})$, the incidence value of this imprecision or tolerance will be delayed on the frequency value in the form of ½ $\Delta C/C$.

Besides imprecisions and tolerances below, it is necessary to take into account the variations of the capacity value in temperature, humidity, etc. (example in automobile, for a contactless starting badge/key before operating correctly from –50 to +75 °C) and sometimes of usual drifts (aging drifts, etc.).

5.3.2.1.3. Capacity electric performances

Of course, during their choices and the design of a contactless system, sooner or later we are interested in the component performances and, for a capacity, the first two that come to mind are the series resistance value (or parallel) and the derived parameter which is the Q quality coefficient of this capacity.

Q quality coefficient of the capacity

This characteristic parameter of the tuned circuits merits a particular section as its implication in the contactless equipment is underlying and discussions and measurements are frequent.

Any L, C-tuned circuit possesses a particular Q quality coefficient as a result of the L inductance is never perfect and possesses its own physical resistance R. Technologically, it is often to obtain values of hundreds or more. At the same time, near field communication (NFC) or radio frequency identification (RFID) applications support each other and are generally governed, on one hand, by standards and protocols having numerical bit rates and known coding bits, collision management procedures, carriers and subcarriers, etc., and on the other hand, they must satisfy the regulatory moments in force of radiation and constraining electromagnetic pollution (in Europe ERO-ERC 70 03, in the United States FCC 47 part 15, etc.). The aforementioned numerical bit rates as well as the form of the signals imposed by the standards (rise time, fall time, cutoff time, etc.) implies fixed bandwidths sizes. Therefore, nearly obligatory values of the application values of the Q quality coefficients of the R, L, C circuits. Often, the latter are brought back to values of Q = ~ 20 with the help of exterior components for equipment NFC (2–5 cm), ISO 14443 cards of the proximity applications type (some cm to 15 cm) and of Q = ~ 70 – 80, rarely above, for RFID ISO 18000-2 (125 kHz) or – 3 (13.56 MHz) or ISO 15693 Vicinity (50–70 cm) applications or other propriety standards RFID (Hitag, Legic, EM, etc.) whose bit rates are weaker than that of ISO 14 443.

Two particular points specific to the "contactless" applications will be highlighted:

– Generally, the very good quality coefficient, Q (1,000 and above), specific to the additional capacities is strongly "surpassed" by the external realization of the Q value (20–80) required by the application.

– Furthermore, the value of the application Q quality coefficient (20–30) is concretely required and useful by the application only at the moment (certainly important) of the starting threshold of the integrated circuit activity (*threshold*) during the presence of weak magnetic fields. After, the integrated circuit consuming more due to the fact that its application (voluntary debit change, implementation of the crypto, etc.) and the conventional voltage shunt regulation internal circuits implemented when strong fields only greatly reduce the Q application value.

5.3.2.1.4. Performance measurements

There are numerous types of measurements related to the contactless equipment particularly the value of the tuning frequency, the Q quality coefficient, the width of the bandwidth, the retro-modulation voltage, etc. (see book PARET 2 constraints on measurements and tests). Some are easy to measure, others are more delicate. Sometimes, some are obtained and derived from measurements of other signals (for example, starting from rising and falling time, etc.). It is also important to note that some types of measurements (particularly those of the proprietary specifications such as EMVCo and NFC Forum) only give views within certain important physical phenomena to be measured (magnetic field values, loading effect and retro-modulation). Whatever it be, when we want to conform to a standard (for example, ISO 14 443 on smart card or 15693 and 18000-3 on RFID or ISO 18092 and 21481 for the NFC), it is necessary to admit defeat of the so-called "Tests Conformance" standards associated with these standards of which the most famous and specific is the ISO 10373-6 standard (350 pages in the 3rd edition 2015) … in which all is described very well.

Example: retro-modulation measurement

In order to help us understand better, we will emphasize a little more on the retro-modulation voltage measurements by variation of passive charge passive load modulation (PLM) which are often subjected to wrong and long controversies when arise from the choice of the tuning frequency of the card. Therefore, from the choice of the capacity value.

Briefly, retro-modulation is the physical phenomenon that allows the target/tag to be understood by the initiator while carrying out a (passive) modulation of its charge (*load modulation*) with the help of a particular coding bit (that is MCS – *Manchester Coded Subcarrier* – or binary phase-shift keying (BPSK)) thus, creating the presence of two frequency subcarriers of strictly symmetrical frequency values of the carrier (at +/– 848 kHz of the latter). This is the principle; see example in Figure 5.2.

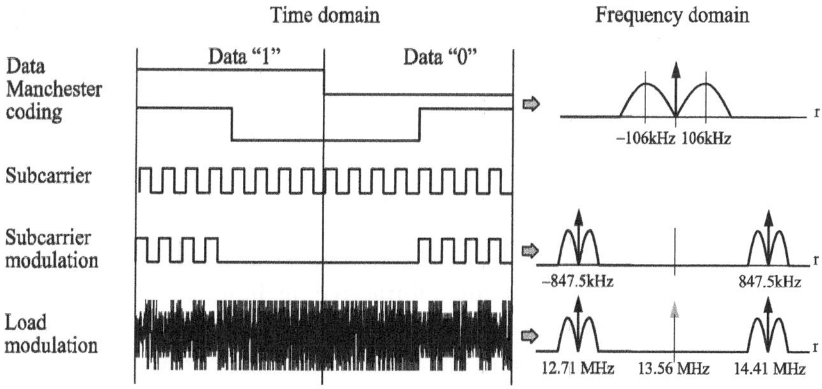

Figure 5.2. *Position of the subcarriers*

The reality is very different, because a particular choice of the voluntary offset of the tuning frequency of the card (see above) the frequencies of these two subcarriers remain strictly symmetrical from the carrier but become completely different in amplitudes, one large, and the other small. At the initiator antenna, with the synchronous demodulator permitting the detection and comprehension of the message contained in this modulation integrates the energy contained in the two subcarriers and it is this entire energy that must be measured. This is why the ISO 10373 – 6 standard indicates the use of a specific measurement method of this energy with the help of a sequence of samples and then of a digitization of the calculation signal follow-up of a DFT – discrete Fourier transform (and not an fast fourier transform (FFT)), chained by a Bartlett triangular windowing function and finally a detection with the help of a Hilbert transform, and not a simple measurement of the peak-to-peak amplitudes of signals such as certain proprietary documents demand (EMVCo and NFC Forum).

5.3.2.2. *Application fields and uses of additional capacities*

As we have already indicated, the technical applications of the contactless field at 13.56 MHz are very varied and diversified. It mainly covers:

– the targets/tags/smart cards well-known conventional contactless ID1 format (bank, of payment, of transport (for example, Calypso/Navigo)) respond to some well-specified proprietary international norms and standards;

– the classic dual cards with contact and contactless and/or equipped with boosted antenna technology booster;

– the access smart card (building (for example, Vigik), office, parkings, skiing area, etc.), the loyalty cards, etc., whatever is in conformity or not to the ISO standards or to other proprietary standards based on manufacturers;

– contactless SIM cards, etc.;

– and finally all the NFC applications and a lot of RFID ones.

For reasons of production costs, the antennas of these targets/tags (often manufactured with wire or with subtractive techniques of graved/etched copper/aluminum or in additive techniques with deposited inks, etc.) have a tendency to reduce in size, and therefore lower their inductance values L requiring in exchange an increase in the capacity value for re-establishing the LC circuit tuning. Furthermore, there are tens of other NFC contactless applications whose mechanical sizes of the final products are reduced and that have very varied factor forms. For example, we mention the “fobs” (key rings), badges, watches, sport bracelets, tags, entry or transport tickets, USB key, tags implanted in baby rats for medical follow-up, on ants and bees, of the tablets, of the objects (device), photo/tablet frame, in brief an inventory like which necessitates mechanically reduced antennas, therefore, weaker L inductance values and higher tuning capacity values than those found in conventional systems.

Table 5.6 recapitulates some examples of common uses at 13.56 MHz in which the use of an additional capacity is justified so as to fill the functionality of the product.

Uses	L	C	Integrated circuit	Additional C	Observations
	μH	pF	Capacity	pF (in E12 series)	
Tag RFID/NFC ISO 15 693 ISO 18 000-3 an other standards	9.2	15	15–20	Without	The tags are cheaper to support the cost of an additional layer
Smart card classic and dual Class 1 & 2	4.6	30	20–21 pF	Without	Cards are tuned to 15/19 MHz
Smart card classic and dual antennas of reduced sizes Class 3	2.3	60	50–70 pF	Without	In accordance with the cost of the integrated circuit
			20–21 pf	47–68	Facilitates more flexibility in the choice of the chip supplier
RFID reduced antenna	1.6	90	90 pF	Without	Example: circuit I-Code NFC IP2 Cost problem
Antenna card of reduced sizes Class 3 & 4 NFC devices	1.4	100	20–21 pF	82	Obligatory additional capacity
Dual cards "Booster" antenna Class 4			50–70 pF	47–27	Very little space for putting the additional capacity
SIM format antenna Class 5 & 6 NFC devices	0.9	160	20 70	120–150 68	Obligatory additional capacity
μSIM format card NFC devices Class 6	0.35	400	20	390	Obligatory additional capacity

Table 5.6. *Examples of common uses at 13.56 MHz of an additional capacity*

5.3.2.2.1. Economic aspects

In Table 5.6, we have indicated the opportunities of the use of additional capacities in NFC under the cover of an attractive economic aspect. Here too, there are different aspects. In conclusion, we take the numerous technological possibilities listed at the beginning of this general introduction.

Capacities integrated in the chip

The solutions in which the capacities are integrated in the integrated circuit with (pre)defined values (often of 20, 50, 70, 90 pF according to the types, associated markets and manufacturers) are in principle the most economical. This can only be defended if the values are sufficient, if the solutions "high capacitance" (70 and 90 pF) have prices that are quite similar to the "low capacitance" solutions and if the latter have high capacitance variants, which are not guaranteed because market segment functions underlie each other.

Completely discrete capacities

In solutions in which the capacities are totally discrete and external in the integrated circuits (conventional components, serigraph technology, etc.) and having, therefore, a large flexibility and liberty of design, we must remember that the integrated circuit only has two connection pads which are generally occupied by the link to the antenna and must also be connected. It is possible to do "*bonding*" (direct or by wire) between chip and additional capacity. However, this presents a supplementary cost which must be compensated by the application.

Juxtaposition of integrated capacities and external additional capacities

These solutions by juxtaposition of integrated capacities and discrete or serigraphed external additional capacities are used when we are not able to apply a previously listed solution and in case we want to reduce the sizes and mechanical volume of the exterior capacity when maximizing the value of the internal capacity. Here, the functionality is more important than the cost.

5.4. State-of-the-art of the antenna sizes

We begin by taking stock of places and art:

– of ISO standards concerning the sizes and formats of antennas directly;

– classes of antennas;

– some secrets of the standard EMVCo.

5.4.1. *Sizes of the target antennas*

This applies to non-normalized sizes, from very large (example, each photo, tablet, etc.) to very small (tag for identification of baby rats) when passing through formats of normalized sizes either by ISO or by the NFC Forum. We begin by the latter, which are quite common on the market.

5.4.1.1. *ISO 14 443 antenna classes*

For cards known as proximity, the ISO 14443 standard organized the possible antenna sizes for PICC (target) in "classes", from larger formats ID1 Class 1 to smaller ones, mini card style SIM Class 6.

VERY IMPORTANT OBSERVATION.– Only equipment whose antennas are in accordance with the formats of the six classes described in amendment 1 of the ISO1 4443-1 standard (see Figure 5.3) can be called, sold and approved under the name "ISO 14443 of class X".

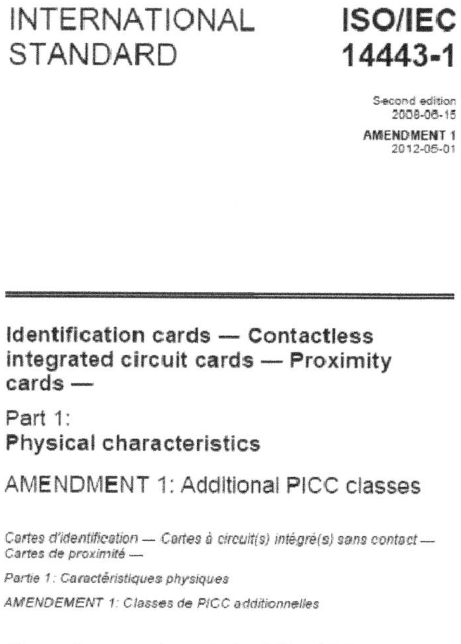

INTERNATIONAL STANDARD

ISO/IEC 14443-1

Second edition
2008-06-15

AMENDMENT 1
2012-05-01

Identification cards — Contactless integrated circuit cards — Proximity cards —

Part 1:
Physical characteristics

AMENDMENT 1: Additional PICC classes

Cartes d'identification — Cartes à circuit(s) intégré(s) sans contact — Cartes de proximité —

Partie 1: Caractéristiques physiques

AMENDEMENT 1: Classes de PICC additionnelles

Figure 5.3. *Fax of amendment 1 of the ISO1 4443-1 standard*

In the drawings, Figures 5.4(a) and (b):

– the *non-shaded* areas correspond to surfaces/areas left free for arranging the antenna winding;

– the *shaded* areas correspond to surfaces/areas which must remain free from all components (coil, additional capacities, etc.) for mechanical reasons.

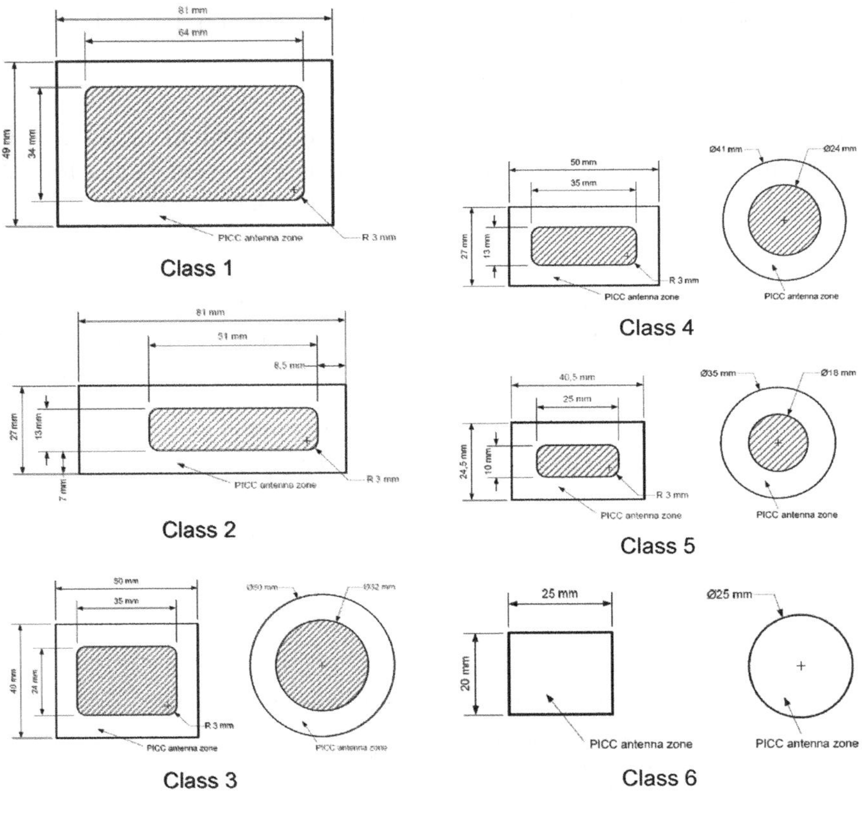

Figure 5.4. *a) Class 1 to 3 mandatory and b) class 4 to 6 options*

5.4.2. Examples of applications of targets with antennas in ISO classes

We give some concrete examples of application fields of these antenna formats in NFC equipment.

5.4.2.1. *Class 1*

These are antennas whose form and surface fill the mechanical format ID 1 or universal integrated circuit card (UICC) of 85.60 × 53.98 mm of the standard smart card, office access badges, visitor cards. It is a great standard. These NFC devices normally have no batteries but are sometimes also of the *battery-assisted* type (with very thin "paper batteries") because they sometimes posses a small local display unit and its associated electronics.

5.4.2.2. *Class 2*

Despite the use of cards, badges, tags or targets in the mechanical format ID1, it is often impossible to be able to use all the surface of class 1 to install the antenna due to exterior mechanical reasons (embossing, fixation holes, etc., expected on the target).

5.4.2.3. *Classes 3, 4 and 5*

Very numerous industrial applications using these antenna classes.

5.4.2.4. *Class 6*

Class 6 is for small antennas designed in accordance with the mini SIM/2FF (25 × 15 mm) sizes and microSIM/3FF (15 × 12 mm). The cases of uses of these antennas comprise of millions of "wearables" often *battery-assisted* (watches, sport bracelets, health gadgets, etc.) having small antennas and using active load modulation – ALM – and on the edge having small batteries showing little incidence on the loading effect.

Classes from 3 to 6 directly implicating certain conclusions:

– smaller and smaller available surfaces of the antennas;

– straight from B inductions of the weaker magnetic flux picked up;

– higher need of field H.

5.4.2.5. *NFC antenna formats outside ISO 14443*

NFC ISO 18092 or 21481 standards (the real, the only) do not define antenna classes and limit themselves to defining (*form factors*) formats leaving free choice to each one for its specific application of using large formats A4 A3 or all small miniatures by passing through the ISO classes or not such as those practiced on mobile telephone in card emulations mode. But of course, not only ISO 14443 exists in the life of NFC, there are also NFC Forum and EMV which have also defined individual antenna for their test bench.

5.4.2.6. *Associated values of the magnetic H fields*

Due to the fact that, on one hand, there is an implicit reduction of the sizes and values of inductances of these targets and, on the other hand, that the integrated circuits (required functionalities, etc.) are as they are, but on the other hand to succeed in remote power supply the latter correctly, it is necessary to re-adjust the normalized values of the magnetic fields necessary for their operation. Therefore, in accordance with the classes presented above, ISO 14443-2 (see Figure 5.4) defined (new) values of magnetic fields necessary to the proper operation with initiators and targets "pure" ISO.

Let us consider certain sections of this standard in detail.

The proximity coupling device (PCD) (the initiator):

– must support the PICCs (targets) of "Class 1", "Class 2" and "Class 3";

– can eventually support the PICCs of "Class 4", "Class 5" and "Class 6";

– must produce a field of at least Hmin and do not exceed Hmax.

The PICC (the target): if the PICC responds to the requirements of a class, this one must function continuously between the Hmin and Hmax values defined for its category (see Table 5.7). This includes all the requirements of the PICC defined in the standard.

	H_{min}	H_{max}
"Class 1" PICC	1.5 A/m (rms)	7.5 A/m (rms)
"Class 2" PICC	1.5 A/m (rms)	8.5 A/m (rms)
"Class 3" PICC	1.5 A/m (rms)	8.5 A/m (rms)
"Class 4" PICC	2.0 A/m (rms)	12 A/m (rms)
"Class 5" PICC	2.5 A/m (rms)	14 A/m (rms)
"Class 6" PICC	4.5 A/m (rms)	18 A/m (rms)

Table 5.7. *Hmin and Hmax by class category*

5.4.2.6.1. Some complements concerning how to measure these magnetic fields

We also find the same values that are shown above in the standard of associated ISO 10373-6 tests which is the associated twin part of the ISO 14443 standard. Furthermore, to be able to estimate these fields, this one describes a calibration coil. It involves the test "PCD assembly" which consists of a 150 mm diameter coil. Due to this coil and starting from the usual relationship:

$$H(a, r) = \frac{1}{[(1 + a^2)\text{^}3/2]} \text{ x } H(0, r)$$

it is quite possible to derive the field H_d on a point of the space located on the antenna axis (see Chapter 2 for the calculations).

EXAMPLE.– For the proximity coupling device (PCD) antenna diameter of the ISO 10737-6 = 15 cm, we pass from a field of 7.5 A/m to a field of 1.5 A/m, (that is a ratio of 5), when the relationship of the distance between the PCD antenna and the card is approximately 1.4 the PCD radius that is a distance of 1.4 × 7.5 = 10.5 cm. Therefore, a PCD supplying a H(0) of 7.5

A/m magnetic field will be able to read all the ISO cards (since before being sensitive to 1.5 A/m minimum) from 0 cm to about 10.5 cm.

5.4.2.7. *Calculation of the fields and magnetic inductions of a minimum threshold*

This section is one of the most important sections of the chapter. To understand all that, let us turn to electronics.

5.4.2.7.1. Electric representation of the target

Let us take the diagram equivalent of the high frequency (HF) part of a target (see Figure 5.5) in which the Rp resistance appears (the equivalent of parallel Ric with R2 resistance restored to parallel).

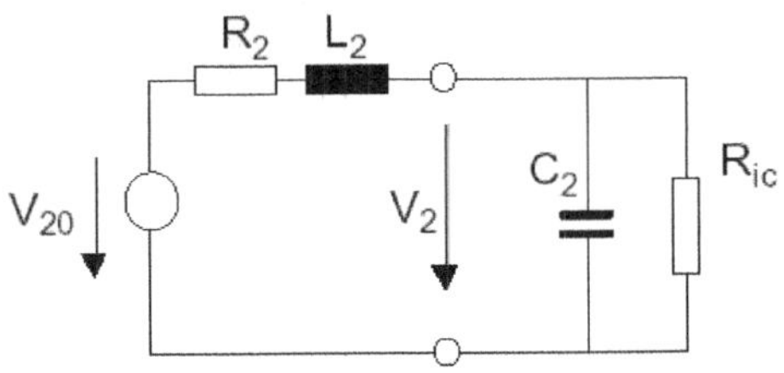

Figure 5.5. *Equivalent diagram of the HF part of a target*

The equivalent diagram of the target "antenna loaded tuned by an integrated circuit" (L equivalent in series with an R and a C parallel) can be represented under its parallel form in which we can call:

L2p = L2s – inductance of the coil/antenna

Cp = C_accord + C_ic + C_con + C2p

= sum:

- from the tuning capacity (plus additional if necessary);

- from the entry capacity of the integrated circuit in the voltage threshold;

- from the parallel capacity of connections;

- of the parallel capacity C2p of the antenna coil.

Rp = (R_ic x R2p)/(R_ic + R2p)

= paralleling of the entry resistance of the R_ic integrated circuit and of the equivalent parallel resistance R2p of the antenna coil.

and L2p x Cp x $\omega\_r^2 = 1$ ω_r pulsation specific to card resonance

and Qp2 = Rp/(L2p x ω_c) ω_c work pulsation (the carrier)

(often, the target is detuned in relation to the carrier and ω_r is different from ω_c)

The general equation giving the v_ic (v2) voltage present at the integrated circuit limits is written as follows (// signifies parallel):

$$v\_ic = \frac{(Rp//Cp)}{Z\_L2p + (Rp//Cp)} \times v20$$

After development and reduction and by replacing the induced voltage by the v20 initiator with its value (– j M ω I1), it comes to:

$$v\_ic = \frac{1}{(1 - L2p\ Cp\ \omega^2) + j\,(L2p\ \omega/Rp)} \times (-\ j\ M\ \omega\ I1)$$

From the value of the complex variable that represents the result of this equation and by asking (L2p Cp ω r²) = 1, we derive the module value of V_ic :

$$V\_ic = \left(\frac{M \times \omega \times I1}{\left[\ \left[1 - \left(\frac{\omega}{\omega r}\right)^2\right]^2 + \left[\frac{L2p \times \omega}{Rp}\right]^2\ \right]^{1/2}}\right)$$

V_ic is proportional to I1 and, knowing one more time that: (M x I1) = N2 x Φ21 = B_d x N2 x s2 = μ x H_d x N2 x s2, it comes to:

$$\text{V\_ic} = \frac{\omega \times (\mu \times \text{H\_d} \times \text{N2} \times \text{s2})}{[[1 - (\frac{\omega}{\omega r})^2]^2 + [\frac{\text{L2p} \times \omega}{\text{Rp}}]^2]^{\wedge}1/2}$$

All things being equal, V_ic is proportional to H_d.

5.4.2.8. *Magnetic field of minimum threshold*

The value of the H_d magnetic field (therefore, for a distance "d") depending on the V_ic voltage (or vice versa) is therefore:

$$\text{H\_d} = \frac{[[1 - (\frac{\omega}{\omega r})^2]^2 + [\frac{\text{L2p} \times \omega}{\text{Rp}}]^2]^{\wedge}1/2}{\omega \times \mu \times \text{N2} \times \text{s2}} \times \text{V\_ic}$$

For the ICs of the target to function correctly and play their role, it is necessary that the V_ic voltage reaches at least the minimum voltage V_ic_min given and guaranteed by the manufacturer, independently or not of his agreement on the value of the carrier frequency originating from the initiator. By introducing Qp2 = L2p ω/Rp in the equation, V_ic_min will be reached for a magnetic field value called threshold H_d_t ("d_t" as threshold at a distance "d") of:

$$\text{H\_d\_t} = \frac{[[1 - (\frac{\omega}{\omega r})^2]^2 + [\frac{1}{\text{Qp2}}]^2]^{\wedge}1/2}{\omega \times \mu \times \text{N2} \times \text{s2}} \times \text{V\_ic\_min}$$

An example of the variations of H_d_t depending on ω (or on f) for a target integrated circuit whose minimum voltage V_ic_min would be imposed is presented again in Figure 6.2, or:

$$H_{\min} = \frac{\min V_{IC}}{\sqrt{2}} \cdot \frac{\sqrt{\left(\frac{\omega \cdot L_{coil}}{R_{IC}} + \omega \cdot R_{coil} \cdot C_{sys}\right)^2 + \left(1 - \omega^2 \cdot L_{coil} + \frac{R_{coil}}{R_{IC}}\right)^2}}{\omega \cdot \mu_0 \cdot A \cdot N}$$

The curve of Figure 5.6 indicates the general look of the H_d_t depending on ω (or on f) for a target integrated circuit whose minimum value V_ic_min would be imposed.

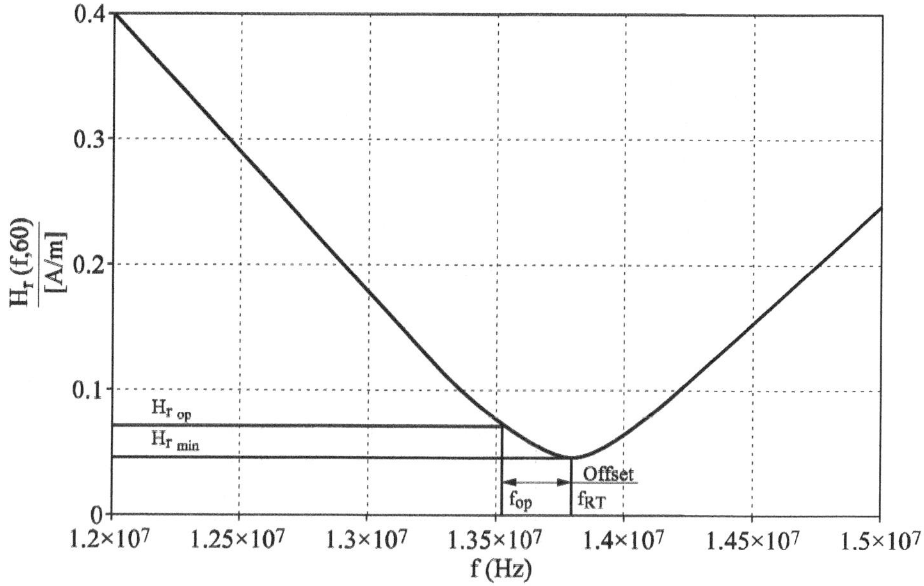

Figure 5.6. *Variations de H_d_t depending on ω (or on f)*

5.4.2.8.1. Specific case or the target is strictly fine tuned

When the target is tuned precisely (ω_c = ω_r) (... *which is rarely precisely the case due to the fact that L and C have discrete values*), and that Qp2 = Rp/(L2p x ω_c), the above equation takes the following reduced form:

$$H\_d\_t = \frac{1}{\mu\_0 \times N2 \times s2 \times Qp2 \times \omega\_c} \times V\_ic$$

With S2 = N2 s2, at 13.56 MHz, this equation becomes:

$$H\_d\_t = \frac{V\_ic}{107 \times S2 \times Qp2} \quad S2 \text{ in } m^2 \rightarrow = \frac{9350 \times V\_ic}{S2 \times Qp2} \quad S2 \text{ in } mm^2$$

Given the hyperbolic form of the H_d_t = f(Qp2) equation, it is a little illusory to increase the value of the quality coefficient Qp2 to hope to reduce the value H_d_t ce which leads to choosing a value of Qp2 adapted to the application that is of 20/30 … and as a result to define the resistance of the antenna coil.

A target obtained according to the staggering of the tuning frequency technique ("*off-center tuning*") of course individually and by principle possesses a value of the threshold magnetic field higher than that of a target tuned on the carrier frequency. For example, all things being equal, a target tuned to 16.95 MHz can require 1.2 A/m against 0.2 A/m to function correctly in relation to the same components tuned strictly on the carrier at 13.56 MHz.

EXAMPLE.– For example, let us examine what will happen when, at the same inductance value of the antenna (L2p = constant therefore Qp2 = constant), we pedagogically voluntarily "detune" the tuning frequency of the latter by modifying the value of the tuning capacity (… therefore in reality, with a given integrated circuit – therefore, at a given entry capacity – we will only be able to adjust the antenna inductance value see Table 5.8).

Target			T0	T1	T2
Frequency of initiator operation	fc	MHz	13.56		
Frequency of target tuning	ft	MHz	13.56	16	19
Relationship "ft/fc"	ft/fc		1	1.18	1.40
Field supplied in d = 0 by the initiator	H_0	A/m eff	4		
Induction supplied in d = 0 by the initiator	B_0	µT eff	5		
Threshold induction for V_ic_min = 4 V eff	B_d_t	nT eff	222	825	1507
(B_0/B_d_t) = (1 + a²)^3/2	B_0/B_d_t		22.5	6	3.3
Hence: a = d/r about	a		2.63	1.65	1.25
Hence operating distance by considering that r = 7.9 cm	d	cm	21	12.6	6.3

Table 5.8. *Influence of the detuning on the field value H_d_t*

5.5. Technological aspect of the NFC targets and tags

5.5.1. *Data specific to integrated circuits for usage by NFC targets*

Each manufacturer of integrated circuits to be used by NFC targets (NXP, Infineon, ST, etc.) has their own noticeably different characteristics (at the same product application ranges). The pertinent electric data of the integrated circuits are only generally indicated in the standard data sheets but given under the *non-disclosure agreement* (NDA). It involves:

– operating threshold voltage (V_threshold);

– input capacity measured by the work frequency and at the threshold voltage;

– input resistance (therefore, power) at the operating threshold voltage;

– input resistance (therefore, power) in operation normal (… to be sure that it will function and will be remotely supplied when, for example, the crypto will start operating).

Likewise, the pertinent mechanical data of the integrated circuit are:

– the silicon sizes of the chip in X and Y;

– the position of the antenna plots/pads (exits) for reflecting to the best connections.

5.5.2. *Data specific to the additional capacities*

When an integrated circuit possesses an integrated input capacity considered too weak to carry out the tuning with the antenna inductance, we can add an external additional capacity to it.

As an example among others, the IPDiA company supplies a range of thin capacities (thickness of 100 μm) of typical values 47–150 pF. The capacities "bumpable" are (quite) easy and integrable on the film/inlay of an antenna.

5.5.3. *Industrial data specific to antenna technology*

This section aims at indicating the principle performances (qualities and limits) of the technologies and can be used for making antennas of the tag and can thereby proceed/help in a choice. This review concerns:

– *the materials*: copper, aluminum, deposited ink, etc.;

– *the mounting technology*: bridge, double face circuit with "via", etc.;

– *the type of antenna*: conventional printed circuit, flexible, plastic leaflet, folded strip, etc.;

– *the associated costs*.

5.5.4. *Technology at stake*

The tag antenna can be executed according to a lot of technology.

Without considering the geometric form (circular, squared, rectangular, etc.) used, the technology of the "inductive" type of antennas can be split into three principal parts which are

– the technology related to the antenna winding;

– the support types (printed circuit, flexible film, etc.);

– the technology linked to the connection techniques of the antenna in the integrated circuit and, if necessary at the connection of an additional tuning capacity of the tag.

5.5.4.1. *Manufacturing technology of the antenna winding*

Several possibilities can be used to constitute the turns of the antenna winding the spools for rolling-up the antenna coil. The inventory of the main technological solutions is often summarized in the list given below.

5.5.4.1.1. Antenna winding with the help of copper or aluminum wires

– Either in the air;

– or on ferrite sticks permitting the concentration of the magnetic field induction lines.

Of course, for economical reasons, it is the first solution which comes to mind. Traditionally, it consists of winding (circular or in a rectangle) copper or aluminum wire. For reasons of costs, winding thickness, mechanical flexibility, etc., a lot of other technologies are also used in the contactless market.

5.5.4.1.2. Etched antennas ("*etched*")

They are generally carried out:

– either on a rigid support of conventional thin printed circuit;

– or on a flexible support (film) facilitating the existence of flexible tags/labels.

At the 13.56 MHz frequency (and in UHF 860–960 MHz) as a result of an interesting quality/price relationship, the main purpose of the industrial market antennas is performed with these two aspects of technology and these must of course be possible to perform in printed or flexible film circuits in accordance with the performances of current industrial tools of the manufacturers which are as follows:

Minimum width of the tracks	*Track*	100–200 μm
Minimum distance between tracks	Gap	100–200 μm
Thickness of the tracks	*Thickness*	35 and 70 μm

Their manufacturing (technically and economically) comes against:

– thickness of copper deposits (costs and difficulties of reproduction);

– gaps (intervals/distances) between turns to avoid the short circuits due to the conductor component smudges (copper, etc.).

A last point that can be taken into consideration is to reduce the loss of the product (copper and aluminum) when support is being carried out: with the engraving technique, if the surface of the antenna network occupies all the support, the weak copper surface occupied by the antenna turns leads to a significant loss of copper during the attack of the printed circuit in the acid.

5.5.4.1.3. Printed antennas performed with the help of a deposit of conductor ink

Instead of extracting copper/aluminum by a chemical process, for (in theory) reducing the costs, we can also deposit conductive ink (a paste + carbon + silver) on a support rigid or flexible insulator (paper, carton, etc.). This deposit can be effected by:

– serigraph, with the help of a deposit of conductive paste … like toothpaste.

Beyond a lower than expected cost, these antennas pose the following problems:

– homogeneous spreading of the ink relief in order to make and guarantee small gaps between coils;

– drying time quite long in order to guarantee good electric and reproducible qualities (ohmic resistance … therefore the coefficient of quality Q) and mechanically stable (extraction of chips), etc.

Often, this antenna technology is used/well adapted in order to carry out cheaper targets/tags whose lifetime is short (ticketing applications, transport, entry into the stadium for sporting events, etc.).

5.5.4.1.4. Antennas carried out by heliogravure or by an offset process

For financial and mass production reasons the antennas are also carried out by heliogravure, offset or vaporization in an aluminum ball vacuum.

5.5.4.2. *Connection technology of the integrated circuit and/or report of the chip*

The technology used to carry out the connection of the antenna winding in the integrated circuit is an essential component for assuring a lifetime correctly adapted to the use of the target. The choice of the technology used for the execution of the antenna winding is also used to choose the technology to be used to ensure the coupling or connections to the integrated circuits. We mention the following techniques for the record.

5.5.4.2.1. Soldering

These techniques are used during the use of delivered chips, not naked but in microcase.

5.5.4.2.2. Bonding

Here, we use connections with the help of microwire of gold (from 10–15 µm in diameter) generally carried out by thermocompression of the wire between the “pad” of the integrated circuit and the printed circuit of the antenna

5.5.4.2.3. Flip chips

To realize the connections between chips and antennas, it is also possible to turn the integrated circuit side against tracks of the printed circuit or antenna ink having taken care before "bumped" the "chips" that is to say having deposited small bumpers/micro drop of gold, on the pads of the integrated circuit of the bumps – in order to be able to, here too, a thermocompression is carried out.

In NFC at 13.56 MHz, the differences between these techniques mainly reside in the mechanical strength to the extraction of the integrated circuit to its antenna.

5.5.4.3. *Technology of the bridge(s)*

In case of using technology of conventional printed circuits or flexible films, it will be necessary to consider the use of "bridges" facilitating the connection of the internal extremity of the concentric spiral that by topological principles constitutes a flat coil antenna on printed circuit. There are many techniques for making these bridges, presented in the following.

5.5.4.3.1. Case of single-sided printed circuit

During the use of single-sided printed circuit, the bridge can be carried out either

– by the integrated circuit itself ... by passing coil tracks under the chip;

– or by passing the antenna above, which requires an insulating layer between the antenna and bridge;

– with the help of an integrated circuit sold in a special microcase making up a bridge office (for example, technology *"Flip Chip"* of type I_code).

5.5.4.3.2. Case of double-sided printed circuit

During the use of a double-sided printed circuit, the bridge can be carried out

– with the help of a via (metallic hole) between the two layers;

– by a via "clip" between the two layers.

5.5.4.3.3. In summary

Briefly non-limitative, Table 5.9 summarizes the main proprieties and presents a comparative between the different technological concepts usable for carrying out tag antennas. Furthermore, numerous companies presented on the market are absolutely capable of solving (or else help in solving) the problems linked with the performance of the inlets.

Antenna technology	Coiled wire	*Embedded*	Engraved	Engraved	Printed
Conductive layer	Copper	Copper	Copper	Copper	Silver conductive ink
Type of interconnection	*Bonding* by thermocompression	*Bonding* by thermocompression	Welding	Adhesion conductor	Adhesion conductor
Quality of interconnection	Very good	Very good	Very good	Average	Average
Breakout force (in cN)	> 200	> 200	> 200	< 60	< 60
Antenna isolation	Yes	Yes	Yes	Option	Option
Ease of handling during the production	Average	Very good	Very good	Good	Good
Quality of the conductive layers	Very good	Very good	Very good	Very good	Good
Possibility of integration of an additional component	No	No	Possible	Possible	Possible
Relative cost in %	110		100	95	85
Comments					Dispersion of the resistance value and of the quality coefficient

Table 5.9. *Comparison between the different usable technological concepts*

5.5.4.3.4. Geometric forms of the winding

Regardless of the execution technology used, their circular geometric forms, squared, rectangular, often take the following aspects:

– in form of flat coils, coiled in wire (whose thickness is very low);

– without thickness (except the thickness of the printed circuit, from the ink deposit, etc.) in spiral, in Greek, etc.

We are now going to quickly examine the majority of these possibilities.

Spreading and the optimal disposition of the tag antennas

The method of spreading and geographical and magnetic arrangement (addition of flow, subtraction of flow, compensation of flow, cancelation of flow, etc.) of the antennas (position as well as the spacing of antennas some in relation to the others) in accordance with the form of the reading area will be detailed during this phase.

5.5.5. *Estimation of the minimum number of antenna coils of the target to guarantee its remote power supply*

Let us consider again the relationship which links the minimum magnetic field value H_d_th to the threshold voltage V_ic_min:

$$H\_d\_th = \frac{\left[\left[1 - \left(\frac{\omega}{\omega r}\right)^2\right]^2 + \left[\frac{1}{Qp2}\right]^2\right]^{1/2}}{\omega \times \mu \times N2 \times s2} \times V\_ic\_min$$

When the target is tuned (ω = ωr), this equation permits the extraction of N2_min to facilitate remote power feeding:

$$N2\_min = \frac{1}{\omega \times \mu \times \times s2 \times Qp2 \times H\_d\_th} \times V\_ic\_min$$

Furthermore, in the above calculations, we have totally ignored the retro-modulation voltage value toward the base station which we must estimate and verify depending on possible coupling between tags and the base station.

6

Detailed Examples of Designs of Target Antennas

In this chapter, we are going to describe the concrete examples of designs to target of target antennas voluntarily chosen in applications having very different form factors from each other in order to give you a look at the largest possible NFC applications and not to limit them only around the mobile telephone.

We begin by the antennas of small sizes.

6.1. Case of small antennas

For reasons of size and/or the reduction of costs, it is often necessary to reduce the size of the antenna. Given that, in reducing the size of the target antenna:

– despite the fact that we add the number of turns, it is difficult to maintain the same inductance value L of the antenna;

– despite the fact that we add the number of turns, it is difficult to maintain the same total surface value of the antenna and collected flow;

– despite the fact that we use a microcontroller having a higher integrated capacity for retrieving or drawing near to the optimal tuning;

– despite the fact that we can be obliged to dispose slim additional capacities to complete the tuning;

...the user wishes to carry out a tag/target close to "Classes 4, 5, 6" while demanding that this should always be in conformity with the EMVCo specifications.

Antenna sizes	Formats Classes antenna	Function of the NFC device		Supply of the NFC device		Retro modulation	Examples of applications (non restrictive)
		Tag	Target	Batteryless	Battery-assisted	LM	
Microscopic	... 11, 12,	×		×			Medical mice follow up Ants follow up
Very small	6		×		×	P/A	Connected objects Mini SIM
		×		×		P	Tag Bracelets
Small	4, 5		×		×	A	Connected objects Watches, bracelets
		×		x		P	Bibs, tags
Standard	1, 2, 3				×	P	Tags, badges
	1.2		×			P/A	Mobile phone in card emulation
Large	A4					P	Bib numbers
Very large	A3	×		x		P	Photo frames
			×		×	P/A	Photo frames

Table 6.1. *Examples of applications based on antenna classes*

6.1.1. *Examples in classes 4, 5, 6... or close by*

Contrary to classes 1, 2, 3 which, according to the ISO standards, are mandatory, classes 4, 5, 6 (see Figure 5.4(b)) are considered to be optional but are required and used increasingly in the applications of tags/targets really NFC. To set the ideas, we mention some of these which are in full expansion.

The first, very fashionable, is that of the invasion of the communicating objects of all types which are mainly in classes 4 and 5 or close by (for example, bracelets for smart watches also an office of card emulation payment, sport and/or (para)medical bracelets etc., whose numerical information, from tracking, etc.[1], is recovered for ulterior processing via an NFC smartphone ... and sometimes up to your health insurance company to be able to conclusively adjust your rights and your tariff).

These targets can be either batteryless or battery-assisted.

The second is that which concerns the struggle against counterfeiting, tracking, applications which are mainly realizable in class 6 or close by (application examples: strong drinks, top class wines, luxury goods (baggage, leather goods, clothes, shoes, under clothes of fine lingerie, designer eyeglasses, etc.), pharmaceutical and drugstore products). In the framework or background of these applications, the bifrequency tags/target UHF (900 MHz) and HF (13.56 MHz), UHF appear for logistical follow-up (long distance – *far field*) during transport, delivery, stocking and short distance – *near field* – control in near field communication (NFC) HF via the NFC ever present smartphone of the truth of the origin of the product, either in a shop by the retailer, or by the final purchaser.

Here are just two small examples of applications that we have just finished seeing, and that is only the start.

6.1.2. *Example of design in class 5*

6.1.2.1. *Connected objects (watch bracelet, etc. or of sport)*

Let us take the example of an NFC target equipped with an antenna of the class 5 type and *battery-assisted* by a small size battery. It involves a connected watch capable of communicating from card emulation mode to active retro-modulation ALM (due to weak coupling between antennas), either with an NFC smartphone, or by carrying out a bank transaction at a payment terminal ((m)POS, DAB) satisfying to the bank qualifications EMVco (reading at 5.5 cm). Furthermore, in order to have better ergonomics of the application, it was decided to increase the size of the antenna to its maximum so as to increase the communication distance by positioning two semi-antennas of 24 × 36 mm in the bracelet on both sides of the watch case

1 For those who have not already guessed, due to connected objects your smartphone becomes the same as an electronic bracelet of prisoners (prisoners of the *Google, Apple, Facebook, Amazon* ("GAFA")) and others to come.

in order to avoid the loading effect of the case (therefore, capable of bearing the name of class 5 ISO with dignity).

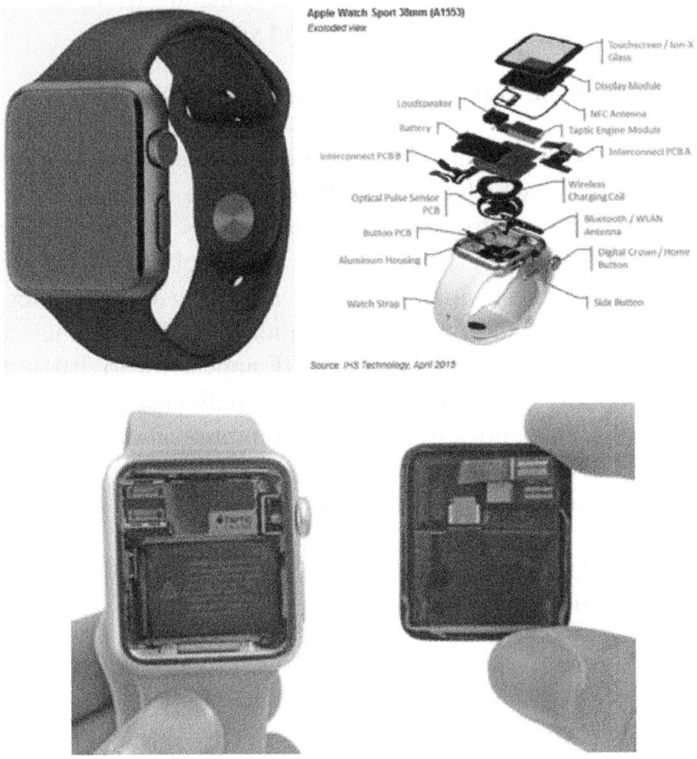

Figure 6.1. *Apple Watch Sport of 38.6 × 33 mm (Doc Teardown iFixit)*

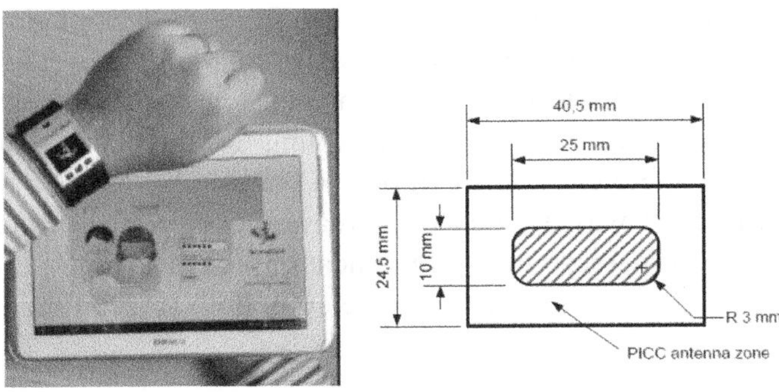

Figure 6.2. *Watch for banking applications*

The chosen integrated circuit microcontroller + NFC can have:

– a maximum size of 3 × 4 mm;

– a particular internal capacity either of 26 pF or 56 pF.

The group can mechanically support an additional capacity of 47 or 100 pF.

The parasite capacity due to support (film + PCV card of permittivity ε_r =~ 2) is supposedly about 5 pF, which is a possible total capacity of about 30, 60, 100 and 160 pF.

For tuning at 13.56 MHz, the following possibilities for realizing the pair L, C are given in Table 6.2.

C	L
pF	μH
30	4.59
60	2.29
100	1.37
160	0.86

Table 6.2. *Tuning possibilities at 13.56 MHz*

IMPORTANT OBSERVATION.– At the same tuning and same exterior sizes of the antenna (therefore, on a mechanical surface called "s"), on the principle it is better:

– to have the smallest total tuning capacity, which leads to the highest L inductance value;

– to have the highest number of antenna turns since L =~ k N^2;

– to have the largest total surface (N × s) collecting the flux.

and therefore the weakest operating threshold field.

In class 5, the mechanical crown remaining for the width of the winding is:

$$(24.5 - 10)/2 = 7.25 \text{ mm}$$

In this placing, the question arising usually is as follows: what are the realizable L inductance values having the maximum number of turns such that the coils made of the total surface collect the most flux possible to ensure the best remote power supply as well as the largest operating distance and if possible that may be compatible with the EMVCo specifications.

We decide to leave 250 μm at the edge of each side of the zone dedicated to the winding, that is 6,750 μm remaining for the widths (tracks) of the turns and the inters turns (gaps):

Track	Gap	Width of a coil + gap	Maximum possible number of turns
μm	μm	μm	
100	100	200	6,750/200 = 33
200	200	400	6,750/400 = 16
400	400	800	6,750/800 = 8

Table 6.3. *Maximum possible number of coils*

In the framework of a class 5, the problem is now, therefore, the following: what are the possibilities of L when the max length of 40.5 mm is given? And what are the unit surfaces "s" and total "S"?

6.1.2.2. *Examples of realization possibilities in class 5*

After some iterations, we obtain the following results for C = 30 and 60 pF.

6.1.2.2.1. Microcase with C = 30 pf therefore L = 4.6 μH

$$S\_tot = 766 \text{ X } 8 = 6128 \text{ mm}^2$$

We can conclude that:

– the number of antenna coils of the tag is of 14 coils;

– the total length of the winding is 17 mm;

– the total width of the winding is 10 mm;

– if the width of the track = ~ gap → 14 + 13 = 27;

– that is a width of track of 3 mm/27 = 110 μm;

– an average length of antenna coil of 17–3 = 14 mm;

– an average width of coil of 10–3 = 7 mm;

– a surface of an average coil "s2" of $(14 \times 7) = 98$ mm^2;

– a total surface "S2" of the tag of $14 \times (14 \times 7) = 1{,}372$ mm² = ~ 13.72 cm^2.

6.1.2.2.2. Microcase with C = 60 pF therefore L = 2.3 μH likewise

$$S\_tot = 885 \text{ X } 5 = 4425 \text{ mm}^2$$

And therefore, finally, here is the great question of confidence so that it works: what are the values necessary for fields and magnetic inductions of minimum threshold?

6.1.2.3. *Calculation of the fields and magnetic inductions of a minimum threshold*

Supposing, for example, that the antenna circuit is (more or less not far from being) tuned to 13.56 MHz, which is often the case, and *Q2 chosen/arbitrarily set at 20 and V_ic_typ = 2 V_eff*:

$$H\_d\_t = \frac{9350 \text{ x } V\_ic}{S2 \text{ x } Qp2} \quad \text{with S2 total surface in mm}^2 \qquad S2 = N \text{ x } s2$$

$$H\_d\_t = \frac{9350 \text{ x } 2}{S2 \text{ x } 20}$$

$$H\_d\_t = \frac{935}{S2}$$

Hence, the following results.

Examples	C Capacity	Sizes antenna of the tag	N Number of turns	S2 Surface antenna total	H_d_t
	pF	mm²		mm^2	A/m
Class 5	30	24 x 40	7	6128	0.15
	60	24 x 40	5	4425	0.211

Table 6.4. *Magnetic fields of minimum threshold*

6.1.2.4. *If that does not happen*

It may be the desired/envisaged solution does not satisfy the ISO and/or EMVCo conformity tests.

If that is not good enough, we increase the Qp2, which is what everybody does if it is possible.

Why? A small movement backward and a stop on image are necessary.

6.1.3. *Example*

The desire is to realize an antenna of 30 × 15 mm with the same integrated circuit without worrying about an ISO class but, as in the preceding case, to verify that this product satisfies EMVCo.

Therefore again, with the Excel Table with S2 = 345.8 × 7 = 2421 mm^2, hence:

Examples	Sizes antenna of the tag	N Number of turns	S2 Surface total antenna	H_d_t
	mm^2		mm^2	A/m
	15 x 30	7	2,421	0.38

Table 6.5. *Magnetic fields of minimum threshold*

6.1.3.1. *Examples of optimization calculations of Qp2 and H_d_threshold*

As we had indicated previously, when the target is tuned to exactly 13.56 MHz this equation becomes:

$$H\_d\_t = \frac{9350 \times V\_ic}{S2 \times Qp2} \quad \text{with S2 total surface in mm}^2 \qquad S2 = N \times s2$$

This equation depends on several variables: N x s2 via S2 and Qp2 that is Ls and Rp (therefore of R_ic and Rs). Yet, whatever happens, the stake consists of reducing the de H_d_t value. For this purpose, it is necessary to maximize S2:

– therefore, S2 max;

– therefore, the largest surface of the coils s2;

– therefore, the maximum N;

– therefore, L is the largest;

– therefore, the integrated circuit has the smallest capacity C so that we can tune the L circuit, C.

It is necessary to maximize Qp2:

– Generally, we cannot maximize the Qp2 value at will because it must have a known maximum value for offering a bandwidth in adequacy with the desired numerical bit rate by the application.

Hence Qp2 max = Rp/(L2p x ω_c) = 15 to 20
and Rp = (R_ic x R2p)/(R_ic + R2p)

that is to say (that in a given integrated circuit, therefore a known C capacity and known L inductance value at the same time) that Qp2, via Rp, is strongly dependent on the entry resistance R_ic of the integrated circuit. Therefore, everything is performed with the ratio value Rp/Lp , that is to say with the choice and performances of the R_ic of the integrated circuit in relation to Lp. Two cases in point, therefore, present themselves:

- *if R_ic is important* (R_ic = 1.5–2.5 kΩ – example: "small" circuits such as Mifare or likewise style), Lp must be quite important to maintain the Qp2 at its max … therefore that the capacity of the integrated circuit be smaller … *hence the absence of additional capacity in that of the integrated circuit*;

- *if R_ic is small* (R_ic = 500–800 Ω – example: "large" circuits, with a lot of memories, of crypto controllers, etc., Smart XM or likewise style), Lp must be weak (but also often at the same S2 reduces) to recover the value of Qp2 … therefore though the integrated circuit capacity be a little more … *hence s' sometimes the necessity to add an additional capacity to the circuit.*

In addition, we can adjust the resistance value R2p a little from the antenna number of turns and redefine it with more precision … that is to say that we can perform/define the thickness (or width) of the track copper of the antenna turns to get the optimal value of Qp2 … hence incidence on the field threshold H_d_th.

Just for confirmation, Table 6.6 gives an example of values.

	Turns			Tracks		Resisfivity		Antenna				IC's			Card		
	number	length	surface	width	thickness	Cu	Al	resistance	inductance	coef Qual	resistance	R_ic	C_ic	V_ic	R2_p	Qp2	H-threshold
	N	l_moy	S2	W	T	ρ_cu	ρ_al	Rs	L	QL_s	Rp	data sheet					
Class			= N s2			17	26	ρNl_moy/WT	Ls=Lp	Ls ω/Rs	=(Q²+1) Rs				Rp//R_ic R2_p/Lpω		
		mm	mm²	µm	µm	Ωm x 10^-9		Ω	µH		Ω	Ω	pF	V_ic_eff	Ω		A/m
5	5.00	70.00	4500.00	120.00	35.00	X	-	1.42	0.86	51.70	3785.89	600.00	160.00	2.00	517.92	7.07	0.588
	7.00	70.00	6000.00	120.00	35.00	X	-	1.98	2.30	98.75	19341.88	2000.00	53.00	2.00	1812.58	9.25	0.337

Table 6.6. *Calculation of the optimal values*

Hence, sometimes recommendations of use such as those below (*taken from* Infinéon).

Table 4 Recommended target card f_{res} [MHz] for various "coil size – product" combinations

Product		M7820	M7892, M7893		M7791, M7794		SLE 97400SD, SLS 32TLC100(M)
Nom. input cap.		27 pF	27 pF	56 pF	27 pF	56 pF	27 pF
Coil size	Class 1	16.5	16.5	-	16.5	-	16.5
	Class 2	14.0	-	14.0	o/-	14.0	-
	Class 3	14.0	-	14.0	o/-	14.0	-
	Class 4	-	-	13.6	-	13.6	-
	Class 5	-	-	13.6	-	13.6	-

Table 6.7. *Recommendations of use*

6.1.4. *Example of design in class 6*

6.1.4.1. *In mini-rectangle (example minicard SIM) (Figure 6.3)*

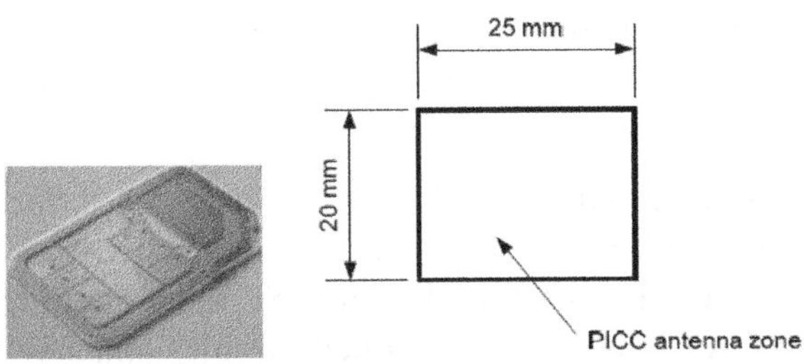

SIM card sizes

SIM card	Reference standard	Length (mm)	Width (mm)	Thickness (mm)
Full size	ISO/IEC 7810: 2003, ID-1	85.60	53.98	0.76
Mini-SIM	ISO/IEC 7810:2003, ID-000	25.00	15.00	0.76

Figure 6.3. *SIM card sizes*

With a microcontroller having a capacity of C = 53 pF, S2 = 7 × 360 mm^2 = 2,520 mm^2 (minicard SIM to eight coils, four of each side and is EMVCo):

$$H\_d\_t = \frac{935}{2520} = 0.37 \text{ A/m}$$

6.1.4.2. *In mini-circle of Φ = 10 mm (Figure 6.4)*

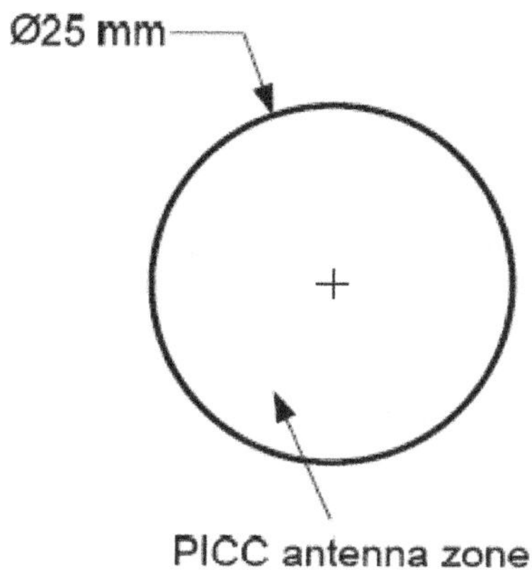

Figure 6.4. *Circular class 6*

With a microcontroller having a capacity of C = 53 pF, S2 = 13 × 43 mm^2 = 560 mm^2, V_thresh = 2 V, Q = 20:

$$H\_d\_t = \frac{935}{560} = 1{,}67 \text{ A/m}$$

With a microcontroller having a capacity of C = 5,353 pF S2 = 7 × 341 mm^2 = 2,387 mm^2:

$$H\_d\_t = \frac{935}{2387} = 0{,}39 \text{ A/m}$$

6.1.4.3. *In mini-circle of Φ = 10 mm (Figure 6.5)*

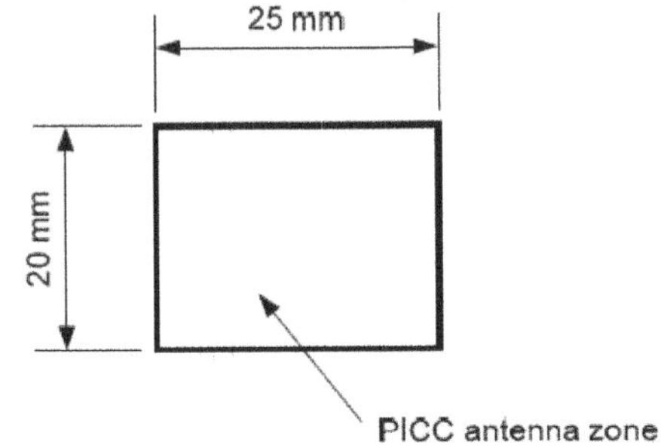

Figure 6.5. *Class 6 rectangular*

With a microcontroller having a capacity of C = 53 pF, S2 = 13 × 50 mm^2 = 650 mm^2:

$$H\_d\_t = \frac{935}{650} = 1{,}438 \text{ A/m}$$

6.1.4.4. *NFC applications for the luxury products anti-counterfeiting struggle*

Often, in the systems which concern the anti-counterfeiting struggle of luxury products, we must ensure:

– on one hand, the logistical tracking of products via a long transmission distance in (RFID) UHF (therefore in a distant field, see Chapter 2 again) or in HF (in a nearby field up to 2 m at 13.56 MHz – see also Chapter 2 – in ISO 15 693) during transport, delivery, storage and warehouse areas;

– on the other hand, HF very short distance (therefore, in a close-field, see Chapter 2 again), controls the verification and genuineness of the origin of the product at the final stage of the in-store sales, either by the retailer, or by the final buyer by using the ever present smartphone as NFC reader.

Given the volume of products at stake, this of course implies that the NFC tags/targets are cheap, therefore, if possible monochip, bifrequency UHF and HF, secured with crypto and, to the choice, of bi-antenna design, either with one in HF and one in UHF or, the "great luxury", a mono-antenna commune UHF/HF.

We present below, as an example, the concept of a concrete realization of an NFC tag equipped with an antenna in class 6.

6.1.4.4.1. Some technology

We begin by some technology. A tag/target antenna can possibly be constituted and considered as the connection of a dipole antenna of a conductor in rings. In this case, this ring is the result of the association of two tuned symmetrical LC circuits.

Let us consider the antenna in circular loops of radius which is represented in Figure 6.6. An I current constantly circulates in the loop which is found in plan xy and is centered on the origin.

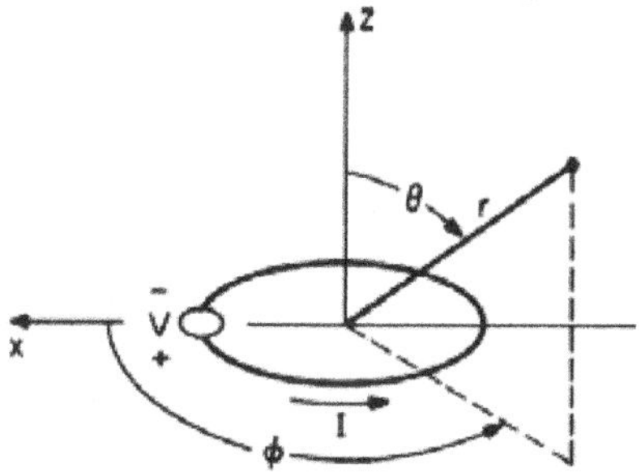

Figure 6.6. *Field of a ring antenna in spherical coordinates*

In this case, the components of a close field of the ring loop (in spherical coordinates) are given classically by the six equations of Maxwell:

$$\left.\begin{aligned} H_r &\approx \frac{a^2 I e^{-jkr}}{2r^3}\cos\theta \\ H_\theta &\approx \frac{a^2 I e^{-jkr}}{4r^3}\sin\theta \\ H_\phi &= E_r = E_\theta = 0 \\ E_\theta &\approx -j\frac{a^2 k\eta_0 I e^{-jkr}}{4r^2}\sin\theta \end{aligned}\right\} \quad kr << 1$$

where k = 2π/λ is the constant of propagation in free space, η0 is the impedance of wave of the free space, r is the distance between the center of loop and the observation point and θ is the angle between the z axis and the direction of the observation point.

From these equations, we notice that Hr and Hθ vary in $1/r^3$ and that Eφ varies in $1/r^2$.

When r is small, the tag/target is found in the vicinity of the antenna of the reader (for example, 10 cm) and it is, therefore, situated in the area of nearby fields of the antenna as a result of that or the components of the field H dominate those of field E and the loop acts as magnetic captor, which increases the sensitivity in proportion to the surface of the loop. The integrated circuit supply is, therefore, mainly due to the magnetic flux through the loop, while the impact of dipole is weak.

For greater distances, when the tag is found in the area of the far fields of the reader antenna, the dipole antenna becomes a captor of a far field increasing the resistance of structure radiation.

A classical realization of this double behavior is shown in Figure 6.7 where the loop is considered as the component in a near field and the dipole is considered as the component in a far field.

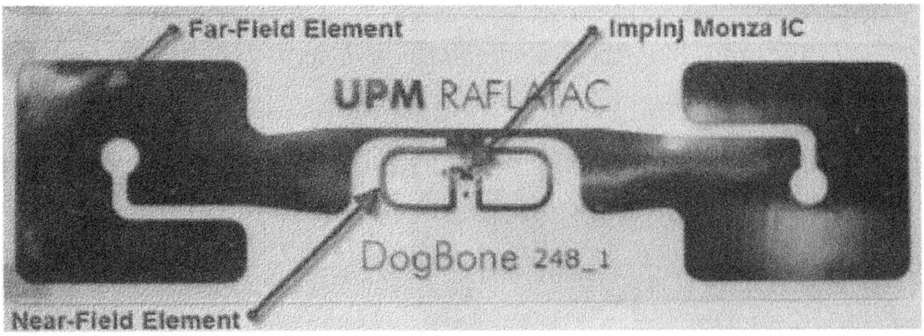

Figure 6.7. *UHF classic antenna operating in far-field and near-field*

The same double behavior that is used above, for example, in a module combining on one hand a small loop (from a single coil *of ~ 10 mm × 10 mm*) and an integrated circuit HF bifrequency for the NFC and UHF for the RFID (Figure 6.8) which, in a standard use with an

NFC smartphone, realized in itself communications at close range of some cm in NFC HF and less than 50 cm with the help of a classic reader UHF RFID (ISO 18000-6C).

Figure 6.8. *Bifrequency (source: Tagsys)*

The distance of reading this module in UHF is a range of 50 cm up to 10 m (for the logistical follow-up) when this module is coupled electro-magnetically in a piece of resonant metal (see Figure 6.9).

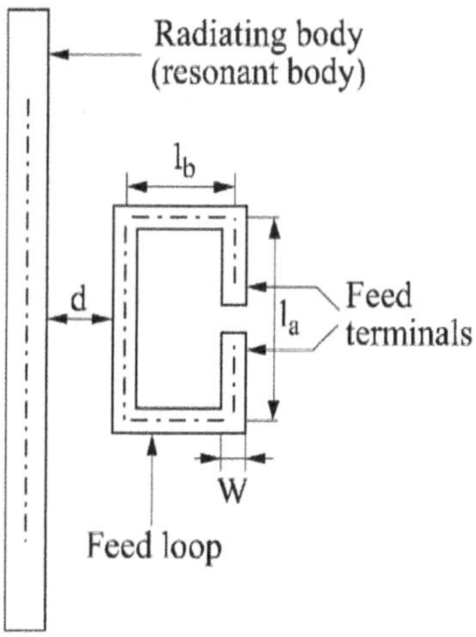

Figure 6.9. *Bifrequency module coupled with a piece of resonant metal*

Examples are presented in Figure 6.10 in the case top of the range shoes with heels of reinforcement radiators, eyeglasses connections or else, for example, for the feminine luxury lingerie in the frames/stretcher of bras.

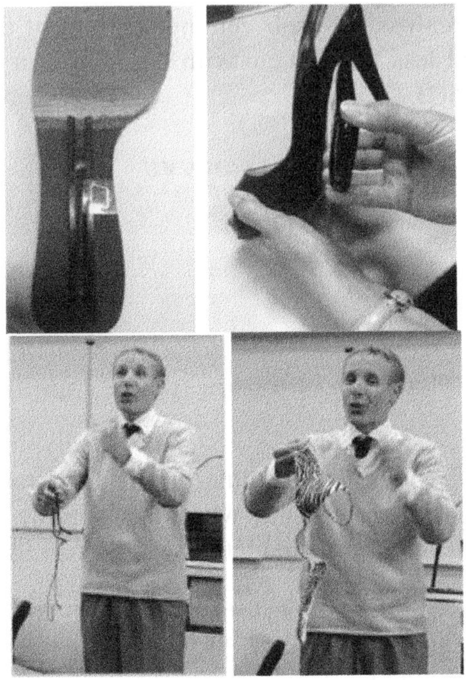

Figure 6.10. *Examples (source: Tagsys)*

As the loop is used as a primary source and not as a specific adjustment component, the performance of the bandwidth can be reduced, but the concept is so flexible and versatile that numerous industrial applications are possible (Figure 6.11).

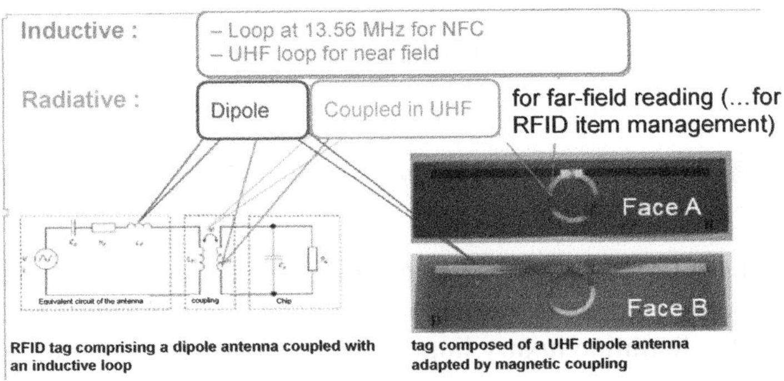

Figure 6.11. *Example of bifrequency concept mono-antenna. For a color version of the figure, see www.iste.co.uk/paret/antenna.zip*

6.2. Case of very small antennas

There are tens of applications requiring very small size targets/tags. Due to their very small sizes, NFC targets/tags can only be solutions:

– mainly batteryless but sometimes *battery-assisted* according to the applications;

– without additional capacity;

– with an integrated circuit having the highest integrated capacity possible for reducing the inductance value to the minimum to find the tuning on the carrier;

– which implies an antenna design on a ferrite stick to succeed in recovering the value of the desired inductance.

Let us give you some example applications and try to imagine the antenna sizes and the final targets/tags:

– detection of military mines during and after the operations (… when we no longer know where we put them– do not laugh, this happens often) with transported tags (stuck) on the back of bees;

– studies and analyses of social behavior, society and hierarchy of ants (in comparison with that of humans) with target/tag ranges by ants (the package of the tag must not exceed 10% of that of the ant in order not to obtain false results of the experiments) (Figure 6.12);

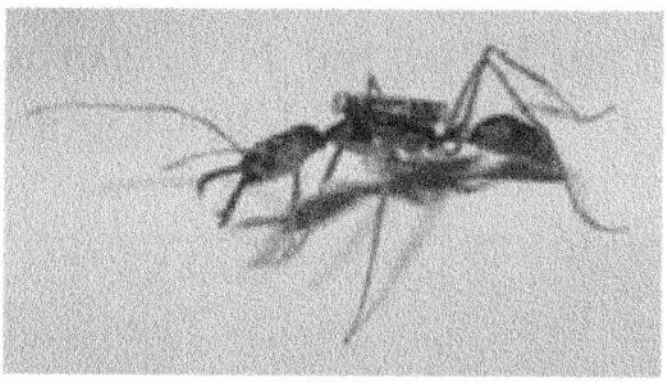

Figure 6.12. *Ant equipped by an NFC target*

– identification, with the help of subcutaneous implants, laboratory baby mice aged 1 day and endowed with a tested DNA and very particular in order not to confuse them and to be able to carry out specific follow-up (of a value of 1,000 € and sometimes more, the baby mouse) (Figure 6.13).

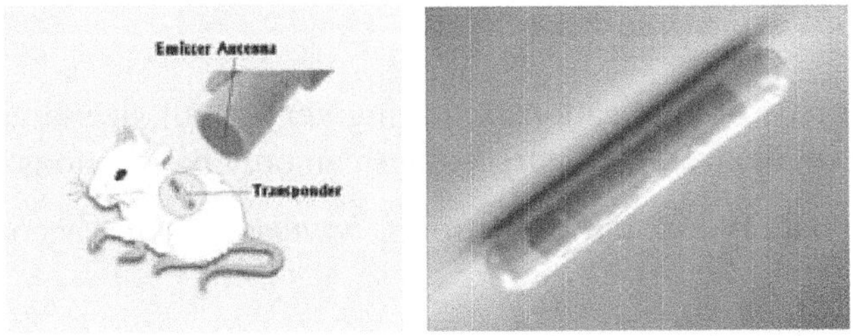

Figure 6.13. *NFC implant for laboratory mice*

Evidently, the reading of all small NFC targets must be able to carry out, if possible, with the help of the NFC mobile telephone standard ... and for mini, microantennas class ISO 11, 12, 13 and following still have not been invented.

6.2.1. *Example of design in classes 11, 12, 13*

Let us take the example of the NFC targets for baby mice[2].

This target is an implant conceived in a glass tube of exterior sizes of 4 mm long and 0.8 mm in diameter in order not to traumatize the unfortunate mouse from its birth. In this tube, there is a ferrite stick with a small flat spot for positioning the integrated circuit and a spiraled winding around the ferrite of very small diameter (some tens of µm). Due to the mechanical and size difficulty of realizing a raised inductance value, the choice of integrated circuit involved a "Tag T2 of the NFC Forum" type at high capacitance value, the MF UC 11 01 "MIFARE Ultra Light High Capacitance" of NXP.

6.2.1.1. *Determination of the inductance value of the "winding + ferrite"*

To determine the theoretical value of the antenna coil inductance, first, "L" does the inventory and the sum "C" of all the participating capacities in the tuning of the coil:

– capacity of the integrated circuit at the voltage of the operating threshold = 50 pF;

---

2 Acknowledgments to Mr Pascal Veillat and Mr Remy Jarretie of the Meusonic, Lutronic, Arelis (brevet WO 2008065127 A2) companies.

– parasite capacity of mounting the chip;

– winding capacity of the coil;

– wiring capacity = total = 5 pF.

and, define/choose – for application reasons – the tuning frequency of the tag. Through the classic formula "Ls C ω^2 = 1", we can, therefore, define what the desired value of "Ls" (coil + ferrite stick) must be:

$$Ls = \frac{1}{C\,\omega^2}$$

that is a theoretical value of 2.5 µH (winding + ferrite)* @ 13.56 MHz.

6.2.1.1.1. Air core coil – Lo

We know that if the coil was a simple air core coil, the value of its inductance would be "Lo" given by the classic formula

$$Lo = \mu\_0 \, \frac{N^2\,Ae}{l} \text{ in Henry}$$

with
- μ_0 = 4 π 10−7
- N = number of circuits
- Ae = section of a coil
- l = length of the winding in the air (do not look for anything else, there is no ferrite)

This formula will facilitate the direct determination of the number of circuits for a given length of winding.

6.2.1.1.2. Coils with ferrite – L

The L inductance value "winding + ferrite" is given by the formula:

$$Ls = \mu\_0 \, (K' \ \mu\_rod') \, \frac{N^2\,Ae}{L\_c} = Lo \times (K' \ \mu\_rod') \text{ in Henry}$$

with:

– μ_0 = 4 π 10−7;

– N = number of circuits;

– Ae = section of the ferrite stick;

– L_c = (L_c for L_coil) length of winding the coil. In order to avoid taking into account the same incidences twice, for the value L_c the physical/mechanical length of the ferrite stick (our case 4 mm) because as indicated below the term "μ_rod" includes the corrective relationship due to the effective length of the stick due to a flat spot for setting up the integrated circuit and "K'" the corrective relationship due to the filling factor.

Besides the term (K' μ_rod'), this formula is strictly that of an air core coil Lo. We must, therefore, introduce the fact:

– on one hand, via μ_rod' = μ_s'_e, the presence of a ferrite stick:

- not particularly cylindrical and homogeneous;

- which only functions especially at low frequency, hence the presence of μ_rod in LF or μ_rod' in HF;

– on the other hand, via K', the fact that the winding of the coils does not necessarily completely cover the whole length of the stick taken for reference of the calculations, hence the presence:

- of K when the coils are contiguous;

- of K' when the latter are not (see below).

– μ_rod' = the permeability value "μ_rod'" represents the value of the permeability μ_r of the block of material (in low frequency) corrected:

- on one hand depending on the operating frequency of the tag with the help of μ_s';

- on the other hand depending on the stick's physical shapes which gives, via the value of Nz, the following sequence:

$$\mu\_r \rightarrow \mu\_s' \rightarrow \mu\_s'\_e = \mu\_rod' = \frac{\mu\_s'}{1 + Nz\,(\mu\_s' - 1)}$$

– K' : K... and, below, its cousin K'... represents a coefficient called "*inductance modifier*" to be taken into consideration in order to estimate the inductance value of the group "winding + ferrite" following the effect of the “filling coefficient” defined as the report between the real winding length “lc” in relation to the effective length of the ferrite stick (“l_e”).

OBSERVATION 6.1.– In relation to the K indicated by the curve, marking K’ in the formula keeps us informed of taking into account the spaces between non-contiguous coils in the case of the non-insulated wires or taking into account the thickness of the e-mail for the contiguous coils):

lc/l_e with lc = winding length (if possible centered);

l_e = length (effective) of the stick.

OBSERVATION 6.2.– The value of “lc” is generally determined by winding mechanical reasons.

The filling coefficient value enables us to obtain the “K” coefficient starting from the figure indicated by the ferrite constructor (see Figure 6.14). If the stick is totally filled k = 1.

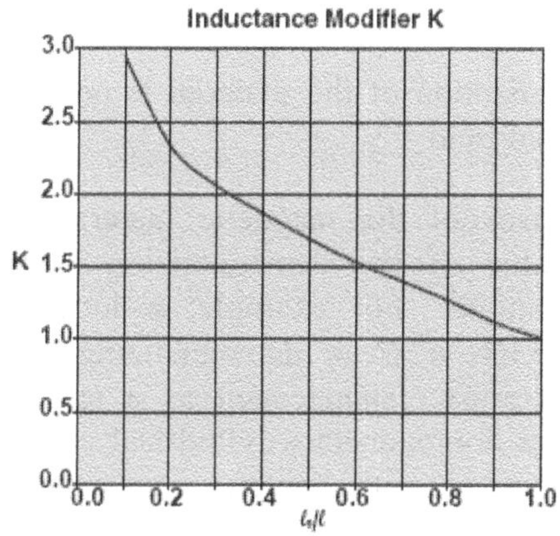

Figure 6.14. *“Modify inductance” K depending on the filling coefficient*

The formula at the beginning of the paragraph facilitates the determination of N for the desired value Ls.

Nz: the vectorial value of field H_d (demagnetizing field) in the stick is equal to:

$$H\_d = -Nz\,\mathcal{M}$$

(the sign "–" indicating the opposite direction of the field) equation in which:

– "$\mathcal{M}$" characterizes the magnetization of the material (in Henry per meter);

– "Nz" is a proportionality coefficient depending on the geometric shape of the ferrite. This term is also called "coefficient of demagnetizing field".

The result is that the magnetic field cleans inside the stick, H_int, is weaker than the dominant exterior field H_ext. In fact, vectorially, we can write that the interior magnetic excitation field is equal to:

$$\overrightarrow{H\_int} = \overrightarrow{H\_ext} + \overrightarrow{H\_d}$$

that is in scalar values:

$$\begin{aligned} H\_int &= H\_ext - H\_d \\ H\_int &= H\_ext - Nz\mathrm{M} \end{aligned}$$

The overall magnetization of the material is not affected except by the reduction of the internal field.

The shape and sizes of the magnetic material have very important incidence on the value of the demagnetizing field _d. This is what denotes the Nz parameter. For example, a long and thin stick of a length/diameter ratio has a weak demagnetizing field, for example, in comparison with a compact shape such as a sphere. For example, we demonstrate that for a homogeneous cylindrical core (stick) of length "l" and diameter "d" (homogeneous diameter on the length of the stick), the value of the Nz quantity can be analytically calculated with the help of the formula:

$$N_z = \frac{d_{max}^2 l}{2\left[l^2 - d_{max}^2\right]^{1.5}} \ln\left[\frac{1+\sqrt{1-\dfrac{d_{max}^2}{l^2}}}{1-\sqrt{1-\dfrac{d_{max}^2}{l^2}}}\right]$$

EXAMPLES.–

– length of stick l = 4 mm unmachined homogeneous stick

$L\_c = 4$ mm
$L\_e = 4{,}87$ mm

– unmachined diameter $d = 0.668$ mm

– equivalent length $l\_e = 4.87$ mm machined stick

– L_e/d $l\_e/d = 4{,}87/0{,}668$
$= 7{,}29$

– we obtain $Nz = 0.052$
$Ae = \pi\, d^2/4$

– permeability/magnetic at f (in MHz) = 0 = 13.56

	f (in MHz) = 0	= 13.56
μ_s'	= 125	= 140
μ_s''	= 0	= 3

hence

$$\mu\_rod' = \frac{\mu\_s'}{1 + Nz\,(\mu\_s' - 1)} = f(\mu\_s' \text{ and } Nz)$$

$$\mu\_rod' = \frac{140}{1 + 0{,}052\,(140 - 1)}$$

$$\mu\_rod' = \frac{140}{7{,}95} = \text{about } 17{,}01 \rightarrow \text{sum of "d } \mu\_rod'/\mu\_rod\text{''"}$$

that is for a ferrite Fair Rite 61 @ f = 13.56 MHz

$$\frac{B\_int}{B\_ext} = \mu\_e = \mu\_rod' = \frac{\mu\_s'}{1 + Nz\,(\mu\_s' - 1)} = 17{,}01$$

IN SUMMARY.–

Ferrite (Fair Rite Company)	f	μ_r	Nz	$\mu\_eff = \mu\_rod'$
				calculated*
Type 61	1 kHz	125	0.052	16.78
	13.56 MHz	140	0.052	17.01

Table 6.8. *Values of μ_e*

For μ_eff'

Taking into account K or K', we can call μ_eff' the coefficient that links the value of Lo to the value of Ls (including winding mode, incidence of the frequency on the losses, etc.).

$$K'\ \mu\_rod' = \mu\_eff'$$

$$\mu\_eff' = K' \times \frac{\mu\_s'}{1 + Nz\,(\mu\_s' - 1)} = f(\mu\_s')$$

Choice of winding

For lc = 1.6 mm (attention! non-contiguous windings)

l = 4 mm

lc/l = 0.4

hence K = 1.8 (in contiguous windings according to the Fair Rite curves)

Therefore "K μ_rod'" = 1.8 × 17.6

"K μ_rod" = 31.7

Value of Ls desired for:

f_0 = 12.39 *f measured with integrated circuit = 12.5*

Σ capa estimated = 50 + 5 pF

→ Ls desired = 2.97 μH

OBSERVATION.– Which means that with the inductance of 2.97 μH, the tuning frequency of the tag will always be a little too low.

$$\omega = 1/\sqrt{(LC)} = 1/\sqrt{(2,97\ 10^{-6} \times 55\ 10^{-12})} \rightarrow f = \text{about } 12.46 \text{ MHz}$$

Calculation of the number of turns

$$Ls = \mu\_0\ (K'\ \mu\_rod')\ \frac{N^2\ Ae}{lc}$$

either:

$$N = \sqrt{\frac{Ls \times lc}{\mu\_0 \times (K' \times \mu\_rod') \times Ae}}$$

$$N = \sqrt{\frac{2,97\ 10^{-6} \times 4\ 10^{-3}}{4\ \pi 10^{-7} \times (31,7) \times (3,14\ (0,668\ 10^{-3})^2/4)}}$$

N = *29 turns*

For information supplements, from the complete equation we "morally" draw the influences of the ferrite stick and the coil position on this one, that is to say we divide its expression by "K' μ_rod" which will give the value of Lo in the air. With Ls = 2.97 μH and "K' μ_rod" = 31.7 representing the influence "ferrite + frequency + winding technique", we obtain:

$$Lo = (2.97\ 10^{-6})/31.7$$

$$Lo = 93.7\ nH$$

In summary, for the antenna coil of the target with a ferrite of the Fair Rite 61 type and in the Terms of Use @ 13.56 MHz, the mechanical characteristic principals, the electrical and magnetic aspect of the tag antenna coil are:

μ_eff	=	17,6	
B_int	=	17,6 x B_ext	
L_ant	=	2,97	μH
N	=	29	turns

R_dc	= 4,5 Ω
R2s	= (Lo μ_s'') ω for a ferrite "61"
Rs	= (93,7 10^-9 x 3) x (6,28 x 13,56 10^6)
	= 23,94 Ω

Q_ant_unloaded	= 8,72 @ 13,56 MHz
F_0_coil	= 41,53 MHz (estimated value)

Tag equivalent electric diagram

With the above components, the tag antenna physically comprises of the components L2s and R2s arranged in a series with a C2p winding placed in parallel on all of them:

L2s = 2.97 μH

R2s = 23.94 ohms

C2p = 5 pF

and of course possess:

– its own resonance frequency: ω = ~. 42 MHz;

– its own quality coefficient: Q2s = ~. 11.

The diagram antenna circuit series can be translated by a parallel equivalent diagram composed of ($Q2s^2 >> 1$):

L2p	= L2s	= 2, 97 μH
R2p	= $Q2s^2$ x R2s	

either:

R2p	= $10{,}57^2$ x 23,94
	= 2674,7 Ω
C2p	= 5 pF

Evidently, for many application reasons (voluntary detuning, connection capacity parasites, detection of targets/tags arranged in a pile – *stack* – specific procedure of collision management, etc.), often the L value is adjusted to bring it to the desired value and/or the capacities are adjusted so as to establish the tuning (or detuning) desired.

6.2.1.2. *Together "antenna circuit of the integrated transponder/ circuit"*

Let us consider the complete diagram of the target composed of antenna and integrated circuit represented under its parallel (Figure 6.15).

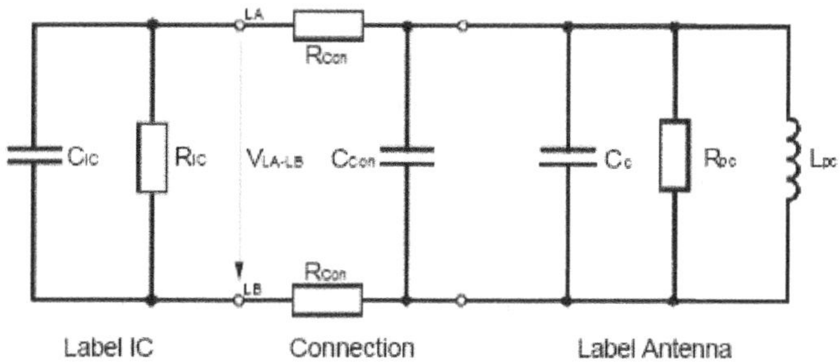

Figure 6.15. *Target complete diagram*

in which we can call:

L2p = L2s

Cp = C_ic + C_con + C2p
= sum of the:
- entry capacity of the integrated circuit;
- parallel capacities of connections;
- parallel capacity of the antenna coil.

Rp = (R_ic x R2p)/(R_ic + R2p)
= paralleling of the entry resistance of the integrated circuit and the equivalent parallel resistance of the antenna coil.

and whose resonance frequency of the whole tag will satisfy the equations L2p x Cp x ω^2 = 1 and Qp2 = Rp/(L2p x ω). At this stage, knowing the value of:

R2p = 2674.7 ohms

and of:

R_ic = 64.5 k ohms

we can calculate, for the whole of "the target antenna and integrated circuit", the Qp2 value–quality coefficient in charge of the whole target. For this, we estimate Rp of the tag:

$$\begin{aligned} Rp &= R2p//R\_ic \\ &= 2674{,}7//64\ 500 \\ Rp &= 2555 \text{ ohms} \end{aligned}$$

which gives a coefficient of overall tag quality:

$$\begin{aligned} Qp2@\ 13{,}56 &= Rp/(L2s \times \omega c) \\ &= 2555/(2{,}97\ 10\text{^-}6\ \times 85{,}157\ 10\text{^}6) \\ Qp2@\ 13{,}56 &= 10{,}12 \end{aligned}$$

IMPORTANT OBSERVATION.– We have shown that for the "coil + ferrite" only:

$Rs = Lo\ \mu\_s''\ \omega$		$= R2s$	$= 23{,}9$
$Ls = Lo\ \mu\_eff'$	with $\mu\_e = 31{,}7$	$= L2s$	$= 2{,}97$
$Qs = Ls\ \omega/Rs$	$= (Lo\ \mu\_eff\ \omega/(Lo\ \mu\_s''\ \omega)$	$= \mu\_eff/\mu\_s''$	$= 10{,}56$

Quality coefficient in charge of the tag – Qp2

Knowing that L2p = L2s, we can write:

$$Qp2@\ 13.56 = Rp/(\ L2s\ \times\ \omega\_c\ )$$

$$= \frac{R\_ic\ \times\ (\ Lo\ \omega\ \mu\_s'^2\ )}{\mu\_s''\ R\_ic\ +\ (\ Lo\ \omega\ \mu\_s'^2\ )} \times \frac{1}{(Lo\ \mu\_s')\ \omega}$$

$$Qp2 = \frac{\mu\_s'\ R\_ic}{(\ R\_ic\ )\ \mu\_s''\ +\ (\ Lo\ \omega\ \mu\_s'^2\ )} = f(R\_ic, Lo, \mu\_s', \mu\_s'', \omega)$$

EXAMPLE.– For a circuit Mifare Light High Capacitance and a ferrite Fair Rite of type 61:

$$\begin{aligned} f &= 13.56 \text{ MHz} \\ \omega &= 85.157\ 10\text{^}6 \text{ rd/s} \\ R\_ic &= 64.5 \text{ kohms} \\ Lo &= 93.7 \text{ nH} \\ \mu\_s' &= 31.7 \end{aligned}$$

μ_s'' = 3

$R\_ic \ \mu\_s'$ = 64.5 10^3 x 31.7
= 204.46 10^4 Ω

R_ic = 6.45 10^4

$Lo \ \omega \ \mu\_s'^2$ = 93.7 10^-9 x 85.157 10^6 x 31.7²
= 0.8018 10^4

$$Qp2 = \frac{204.46}{6.45 \ \mu\_s'' + 0.8} = f(R\_ic, Lo, \mu\_s', \mu\_s'', \omega)$$

EXAMPLE.–

	in BF	μ_s''	= 1	Qp2	= 20,67
Ferrite 61	to 13.56 MHz	μ_s''	= 3	Qp2	= 10,146

Yet, in the application, we are no longer in the air. We must, therefore, take into account the existing relationship between the values of the external and internal magnetic fields/inductions to the winding of the antenna coil. Due to the ferrite (type 61), of its shape (stick), of the operating frequency @ 13.56 MHz:

$$\frac{H\_int}{H\_ext} = \frac{B\_int}{B\_ext} = \mu\_e = \mu\_rod' = \frac{\mu\_s'}{1 + Nz(\mu\_s' - 1)} = 17{,}6$$

Therefore, if external H_d_t is required in the air to produce the V_ic_typ to supply the tag, 17.6 times less is required to produce the same effect for the specific tag (if it is aligned according to the main axis (because of the ferrite)), due to the presence of its ferrite, shape, sizes.

EXAMPLE.–

ω = 2 x 3,14 x 13,56 10^6 s2 = (3,14 (0,668 10^-3)²/4)
μ_0 = 4 x 3,14 10^-7
N2 = 29 Q_ant_unloaded
V_ic_typ = 2,2 eff
Qp2 = 10,12

These values lead, in the air, to the value of H_d_th:

$$H\_d\_th = \frac{1}{\mu\_0 \times N2 \times s2 \times \omega} \times \frac{1}{Qp2} \times (V\_ic\_typ)$$

either:

$$H\_d\_th\_eff = \frac{(2{,}2)}{(4\times\pi \times 10\text{^-}7 \times (29 \times (\pi \times (0{,}668\ 10\text{^-}3)^2/4)) \times (2\times \pi \times 13{,}56\ 10\text{^}6) \times 10{,}12}$$

$$H\_d\_th\_eff = 200 \text{ A/m}\ldots$$

in the air, without the presence of the target and for operating the tag equipped with its ferrite, it is only necessary that the initiator supply – in the air outside the presence of the target – in a point of the principal a field:

$$H = 200/17{,}6$$
$$H\_d\_th\_eff = 11{,}36 \text{ A/m}$$

6.2.1.3. *External/commercial data sheet tag*

In conclusion, the following is how to construct "the technical specification for commercial usage" of the tag.

- target in conformity with the NFC ISO 18 092 & ISO 14 443 – 2 & 3 A standards

- operating frequency	f_c	= 13.56 MHz
- resonance frequency of the tag	f_t_typ	= 12.46 MHz
- quality coefficient of the tag	Q_t_typ	= 10.12 @ 13.56 MHz
- bandwidth of the tag	BW_t_typ	= 1.34 MHz at -3dB

- minimum value of the threshold of magnetic field in the air to be in operating mode

H_d_th_min = 11.36 A/m rms

Box 6.1. *Example of technical specification in commercial usage of the NFC tag*

6.3. Case of the large NFC target/tag antennas: format A4

In order to show that the antennas of the NFC targets offer a very large panel, we now take an example located very, very far from the preceding.

6.3.1. *NFC bib number antennas for marathon and triathlon runners*

In order to present the vast application fields (and of vast problems too) of the NFC targets/tags, here is the example of the bib numbers for marathon (… and triathlon runners, it is more fun, we must take into account the passage in water)[3]. This case was highlighted from 2001 at the beginning of the use of circuits I_code supporting the ISO 15 693 standard and therefore became compliant to "NFC" at the appearance of the NFC IP2 – ISO 21 481 standard. At this period, this application was not yet returned under the name "T2" in the nomenclature of the NFC Forum … which only came into existence in 2004.

The problem of reading bib numbers of the runners at 13.56 MHz is twofold because these must be able to be read:

– on one hand, during the course, at a great distance (2–3 m which, whatever those who are not convinced say, is still a near field distance at 13.56 MHz – see Chapter 1) by gantries whose antennas are about 2 m high by 3 m wide (see Chapter 7 on initiator antenna);

FIG 2

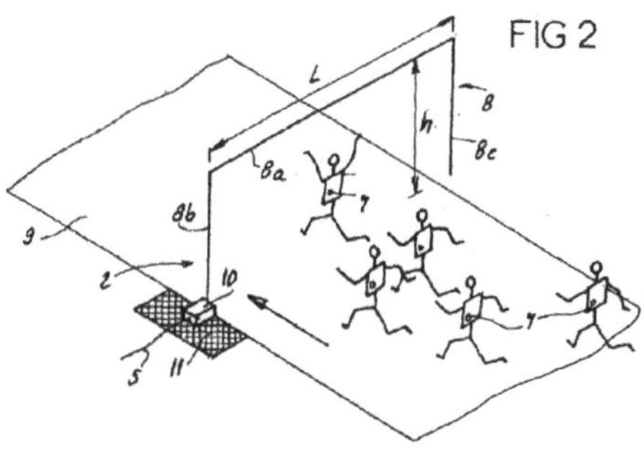

Figure 6.16. *Reading of NFC tags at long distances*

3 Acknowledgments for their distant technical collaboration and friendly to MM Serge Munnia, Tony Bernard and Jean Marc Martin of the Pygmalyon Company – DAG System (PCT 2003 002267 license).

– on the other hand, before departure, at short distances to initialize the content (name of participant, category, etc.) and finally after the arrival to control the passage to the passage points along the trajectory, the non-trickery, etc., via a cheap nearly universal ever present reader, i.e. the NFC smartphone

6.3.2. *Technical properties required by the NFC target/tag*

– Format of the bib number antenna = 0.2 × 0.16 m and a S_max = 0.3 m^2 (0.5 x 0.5 m).

– In order to free the body (e.g. of the transpiration, of the passage in water, ... summary of all that causes strong variations ΔC (increase) of values of tuning and a possible detuning capacity of the target antenna), these effects should be muzzled by having the strongest tuning capacity value if possible (therefore the lowest inductance) by adding exterior additional capacities to the integrated circuit.

– This tuning capacity maximum value is capped by the fact that, during the course, the antenna must also function (remotely supplied and read) at 3 m and collect the maximum flux (... or that which remains) therefore have the largest surface compatible with the mechanical format of the bib numbers, therefore a certain surface, therefore a certain minimum inductance, therefore the coil minimum; that is to say one.

– All these compromises leading to:

- standard bib numbers, antenna of 1 rectangular coil of 160 cm × 140 cm;

- length of the coil = 60 cm;

- L inductance of the antenna calculated at 0.6 μH;

- hence, total maximum tuning capacity at 250 pF;

- furthermore, we must obtain an important Q, for example Q max = 30;

- this reverts to R = Lω/Q = (0.6 × 85.157)/30 = 1.7 ohms.

– The problem, therefore, consists of resolving the problem of the value of the resistance R. Let us calculate the classical value of the continuous antenna resistance (DC):

$R\_dc = (\rho\ N\_ant\ l)/s$

- either with mono-coil antenna in printed circuit with a section of copper of: s = (1.8 10^-8 x 0.6)/1.7 = 0.6 10^-8 m^2;

- or with a realization in printed circuit of 35 µm thick that leads to a track width of: 60 10^-10/35 10^-6 = 170 µm;

– At 13.56 MHz, because of the skin effect at high frequency, a simple wire of copper, which possesses a continuous weak resistance as above, sees its impedance increase (see Chapter 7 for more details on initiator antenna) and this necessitates the use of a wire divided into several thinner strands, enameled and therefore electrically insulated from each other – called "Litz wire" – which facilitates the reduction of these problems. Each of these strands must have an inferior diameter than the value of the skin thickness to the working frequency. Furthermore, due to the proximity and interactions between strands, the current is also not distributed between all the strands and for the current to be distributed equitably, each alternative strand between different position by making tresses or stranding. In our case, after calculations, this can be resolved with the Litz wire with seven strands of diameter 0.25 mm.

– That, again, involves two spaced concentric coils of the adequate distance, which is internal in short-circuit only serving resonator/radiator per re-radiation such as operate *backscattering* of an RFID antenna in short-circuit in UHF (see Dominique Paret's work – *RFID in UHF* at Dunod/Wiley).

6.4. Case of very large antennas targets: format A3

6.4.1. *Context and technical frame of the large antennas*

The objective of this example is to show how to build (a network) of antennas for an NFC device facilitating:

– reading by a photo frame/tablet (the initiator) of the content of a target (a mobile telephone in card emulation mode) on the stretch of the initiator surface;

– or, on the contrary, downloading photos on mobile telephone (the initiator depending on reader writer) toward a photo frame/tablet (the target).

An NFC conventional device does not usually facilitate reading/writing in a concentrated area around its antenna, which can cause problems of ergonomics when the specific delimitation of this area is not possible. Due to the technical concept that we will describe below, the tag reading area can be extended by linking/coupling several identical antennas, according to a network connected to the integrated circuit tag.

6.4.2. *Retained concept*

The classical application is of type "target mode of passive communication", the frame/tablet antenna is remotely supplied by the interrogator (the mobile telephone).

The technical concept chosen to carry out the extension of the target surface area consists of linking several identical antennas between them to form a particular network in order that, given that through the target integrated circuit, the overall inductance value remains equivalent to the inductance of a single antenna and that the magnetic flux necessary to the supply of the tag be sufficient.

One of the ways of resolving this problem is to form an antenna network using parallel n connections made up of n antennas in series (see Figure 6.17). Apart from any magnetic coupling between antennas, the total inductance of the network is reverted to the inductance value of a single antenna. The equivalent inductance seen between points A and B of circuits 1 and 2 remains identical only. This network thus created can be connected directly to the limits of a conventional integrated circuit for NFC tag applications.

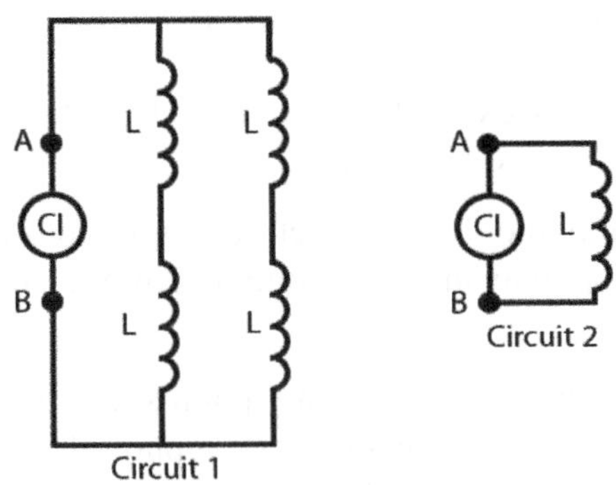

Figure 6.17. *Antenna network of n parallel connections made up of n antennas in series*

In this application, a minimum operating distance of 20 mm is envisaged with the mobile telephone, and to complete the description of this project, it is necessary to arrange parameters given below.

6.4.2.1. *From the manufacturer of the mobile telephone*

– Shape of the mobile antenna = rectangular.

– Mechanical surface of the mobile antenna = 4.5 cm × 11.2 cm = 50.4 cm^2.

– Number of antenna circuits N2 = 3.

– Magnetic field that radiates (H_min in A/m) in the center according to the axis principle of the antenna:

- H_0_front = 1 A/m rms on the side of the screen;
- H_0_back= 0.5 A/m rms from the side of the rear.

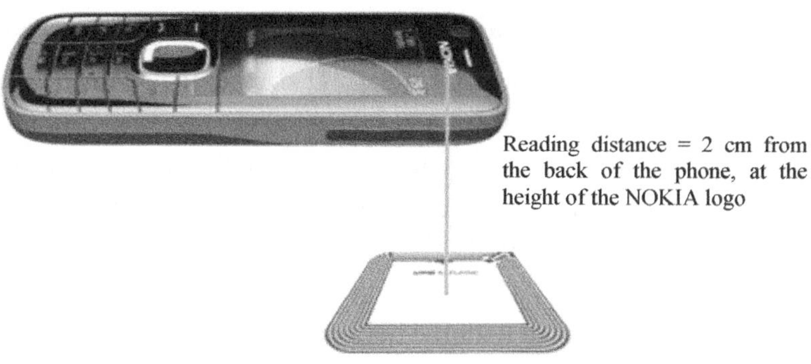

Figure 6.18. *Measurement of the initiator-target distance*

As it can very often be observed, the mobile telephone does not satisfy the NFC ISO NFCIP1 standard which concerns the value of the radiated fields. Of course, it is easy to understand the problem that this is due to compromise mainly bearing on the consumption/autonomy of the mobile, but it only prevents the name "NFC" for an NFC device become a little usurped and close to certain confusion.

*From the manufacturer of the integrated circuit (*tag T1T of the NFC Forum, chip Topaz from Innovision).

The values are as follows: at 13.56 MHz:

– V_ic_threshold_min = 2.14V Vrms;

– V_ic_threshold_max = 2.27V Vrms;

– Input capacitance C_ic @ V_threshold : C2p = 21.1 pF

– I_ic_DC_average_current_consumption_at_V_threshold_max = 200 μA
P_max_ic_power_consumption_at_V_ic_threshold = 500 μW.

6.4.2.2. *From the manufacturer of photo frames/tablets*

– Shape of the photo frame/tablet = rectangular;

– size of the frame = 14.5 × 11.2 cm;

– width/thickness of the frame = 3.5 cm;

and of its immediate electronic/magnetic environment.

Following the provision of all these components, we can now establish:

– the shape (squared, rectangular, etc.) and sizes of the adapted antenna to the mechanical constraints of the photo frame/tablet;

– the choice of the material to be used to realize the antenna (copper, aluminum, conductor ink, etc.) depending on the operating parameters, sizing, production possibilities (industrialization) as well as indications of related costs;

– the tag reading distance with the help of a reference telephone mobile.

6.4.2.3. *Determination of the nominal inductance value "L" of the antenna*

We will use, during the next sections, the following rating.

6.4.2.3.1. For the antenna of the tag

– Antenna series representation:

- Ls_ant: antenna series inductance;

- Rs_ant: antenna series resistance.

– Antenna parallel representation:

- Lp_ant: antenna parallel inductance;

- Rp_ant: antenna parallel resistance.

– Quality coefficient from the antenna to the pulse ω:

$$Q\_ant = \frac{Ls\_ant \times \omega}{Rs\_ant} = \frac{Rp\_ant}{Lp\_ant \times \omega}$$

– Relations between the series and parallel representation:

$$Rp\_ant = Rs\_ant\left(1 + Q\_ant^2\right)$$

$$Lp\_ant = Ls\_ant\left(1 + \frac{1}{Q\_ant^2}\right)$$

If Q_ant >>1, we have Lp_ant = Ls_ant and Rp = Q^2 Rs:

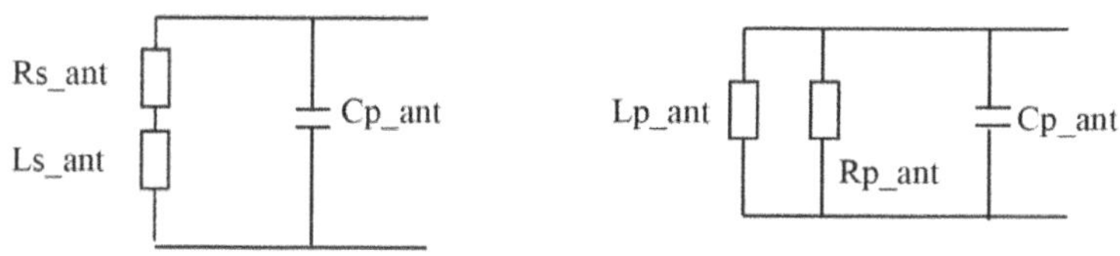

Figure 6.19. *Series and parallel representation*

6.4.2.3.2. For the integrated circuit

– Parallel representation of the integrated circuit:

- Cp_ic: parallel capacity of the integrated circuit;

- Rp_ic: parallel resistance of the integrated circuit.

– Series representation of the integrated circuit:

- Cs_ic: series capacity of the integrated circuit;

- Rs_ic: series resistance of the integrated circuit.

– Quality coefficient from the integrated circuit to the pulse ω:

$$Q\_ic = \frac{1}{Rs\_ic \times Cs\_ic \times \omega} = Rp\_ic \times Cp\_ic \times \omega$$

– Relations between the series and parallel representation:

$$Rp\_ic = Rs\_ic\left(1+Q\_ic^2\right)$$

$$Cp\_ic = Cs\_ic\frac{1}{1+ \frac{1}{Q\_ic^2}}$$

If Q_ic >> 1, we have Cp_ic = Cs_ic.

By having different components obtained from the integrated circuit manufacturer, we can derive:

– resistance value of equivalent parallel entry of the integrated circuit, either:

P_ic_max	= V_ic_th_max ^{2}/R_ic_min
449 10^-6	= 2.27^2/R_ic_min

either:

Rp_ic_min	= 11,476 kohms

– value of the integrated circuit quality coefficient:

Q_ic	= Rp_ic x C_ic x w
	= 11,476 10^3 x 21,1 10^-12 x 85,157 10^6
	= 20,62

– value of the integrated circuit equivalent series resistance:

Rs_ic	= Rp_ic_min/Q_ic^2
	= 11476/20.62^2 = 27 ohms

If we want to transfer the antenna energy maximum of the tag to the integrated circuit that implies the conjugated adjustment of the coil and integrated circuit impedances, that is to say:

1) That the tuning (resonance) of the tag is carried out: to determine the inductance "L" value of the antenna coil, first the inventory and the sum "C" must be done from all the capacities participating in the coil tuning:

– capacity of the integrated circuit; at the voltage of the operating threshold;

– parasite capacity of mounting the chip;

– coil winding capacity;

– wiring capacity.

and define/chose – for application reasons – tuning frequency of the tag. Through the classic formula "$L\ C\ \omega\_tag^2 = 1$", we can, therefore, define what the desired value of "L_ant" must be:

$$L\_ant = \frac{1}{C\ \omega\_tag^2}$$

At 13.56 MHz:

$$\begin{aligned} \mathbf{C} &= C\_ic + C\_bobine + C\_parasite \\ &= 21.1 + \textit{tuned } 4.9\ pF \end{aligned}$$

$$\begin{aligned} L\_ant &= 1/((85{,}157)^2\ 10\text{^}{+12})\ x\ 26\ 10\text{^}{-12}) \quad \ldots \text{ tuned} \\ &= 5.3\ \mu H \end{aligned}$$

2) That the tag antenna resistance values and Rs_ic are equal:

$$\begin{aligned} Rs\_ant &= Rs\_ic \\ &= 27 \text{ ohms… max value} \end{aligned}$$

3) Which would lead to a Q_tag (often called Q2p) in equal charge to half of the Q_ic:

$$\begin{aligned} Q2p &= 20.62/2 \\ &= 10.31 \end{aligned}$$

4) That is a bandwidth at – 3 dB of:

$$\begin{aligned} BW &= f\_c/Q2p \\ &= 13.56\ MHz/10.31 \\ &= 1.315\ MHz \end{aligned}$$

6.4.2.4. *Parasite capacity of the antenna coil*

The coils of the winding consisting of the antenna coil, insulating support, etc., imparting to it a sum of parasite capacities by construction.

Without any measurement, the value was estimated at about C2p = 4.9 pF (… because that provided for the calculations and the order of magnitude is not false), which with an inductance value of L2s = 5.3 µH would give a natural frequency value of resonance of:

f = from 35 to 40 MHz

For information usually, during the realization of the first step model, with the help of a network analyser, we measure the resonance natural frequency of the antenna coil in order to confirm the estimation of the parasite capacity value.

6.4.2.5. *Network of antennas*

It is necessary to evaluate that the value of the voltage will be available at the terminals of the entries of the integrated circuit in the presence of a network consisting of x antennas arranged as indicated in the initial diagram in “parallel series”.

OBSERVATION 6.1.– The parasite capacities of the antenna, wiring, etc., can be reverted parallel on the capacity of the integrated circuit for giving the total capacity Cp_tot.

OBSERVATION 6.2.– V_ind is the sinusoïdal voltage induced by the product magnetic field by the mobile telephone on one of the network antennas. Its value is equal to:

$$V\_ind = -\frac{d\Theta}{dt} = -\frac{\partial}{\partial t}(B\_d \times S\_ant)$$

equation in which S_ant is the total surface of the tag antenna (S_ant = N_ant x s_ant) and B_d is the magnetic induction at the distance “d” of the mobile telephone antenna.

The magnetic field radiated by the mobile telephone is a sinusoïdal field pulse ω_c. The induced on the tag antenna will, therefore, be a sinusoïdal voltage of the same pulse that it emits by the mobile telephone. We can use the complex ratings, either :

V_ind = - j ω_c x B_d x S_ant

OBSERVATION 6.3.– Despite the heaviness that you will sometimes note, we have voluntarily left all the literal equations in an integral condition in order that you can, subsequently, simplify them depending on your specific usage.

6.4.3. *Example of network with four antennas*

6.4.3.1. *Hypothesis no. 1*

The antennas are sufficiently separated from each other (for example, in the four corners of the frame or tablet) and the magnetic coupling between the initiator (the mobile) and the frame (target) only works through a single antenna (see Figure 6.20).

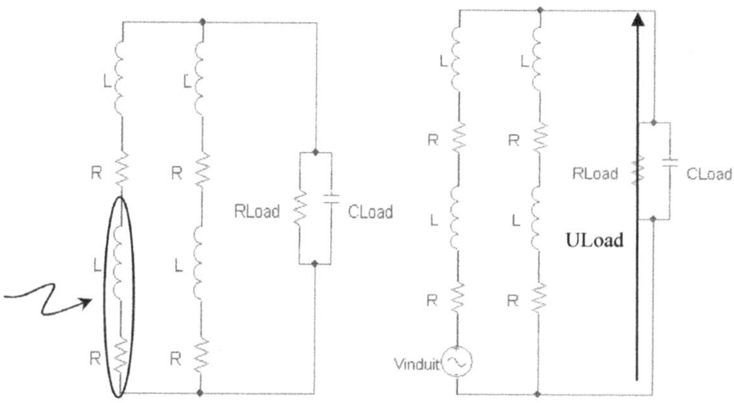

Figure 6.20. *A single illuminated antenna*

After the calculation of the impedance value equivalent to the charge which represents the integrated circuit, then the calculation of the impedance value resulting from making the charge parallel (integrated circuit) with the network connection not receiving the flux participation originating from the initiator, we end by calculating the voltage value which develops in the terminals of the integrated circuit V_ic, in the presence of an induced voltage V_induced following the presence of a magnetic field in the winding of one of the target coils, that is:

$$\frac{V\_ic}{V\_ind} = \frac{Rp\_ic}{2Rp\_ic + 2(Rs\_ant + jLs\_ant\omega\_c) \times (1 + jRp\_ic \times Cp\_tot \times \omega\_c)}$$

6.4.3.2. *Hypothesis no. 2*

The coupling between the initiator (the mobile) and the antenna network of the frame (target) works and is divided into two antennas. In this case, two subcases of the figure must be considered:

– the antennas belong to the same connection of the network;

– the antennas belong to the different connections of the network.

6.4.3.2.1. The two antennas belong to the same connection

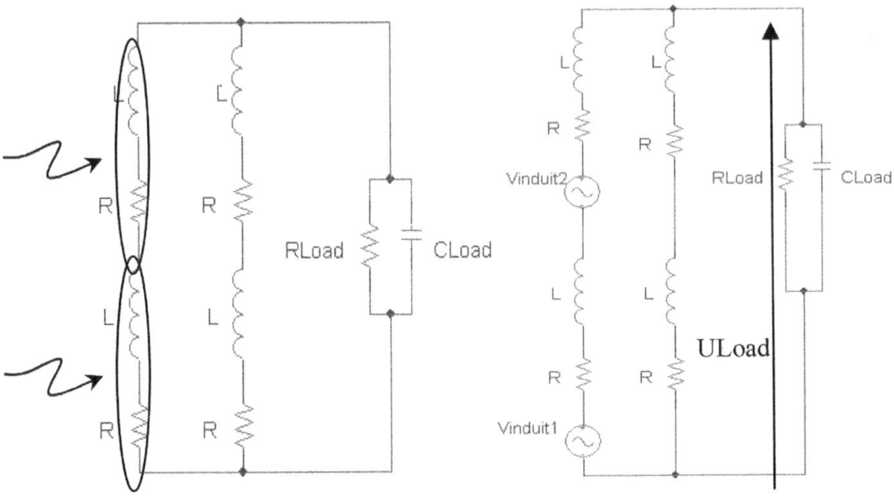

Figure 6.21. *Two antennas of the same connection are illuminated*

By using the superposition theorem, the following result is obtained:

$$V\_ic = \frac{Rp\_ic}{2Rp\_ic + 2(Rs\_ant + jLs\_ant\omega\_c) \times (1 + jRp\_ic \times Cp\_tot\omega\_c)}(V\_ind_1 + V\_ind_2)$$

OBSERVATION.– The induced voltages 1 and 2 indicated on the diagrams can have several directions:

– as indicated above, the magnetic flow emitted by the initiator supplies two antennas with each receiving a part of the flux;

– or else by coupling from nearby (from coefficient of coupling "k"), an antenna re-radiates on another. It may not be the case due to the fact that the antennas will essentially be coplanars and the coupling coefficients between antenna will be of low values.

6.4.3.2.2. The two antennas belong to two different connections

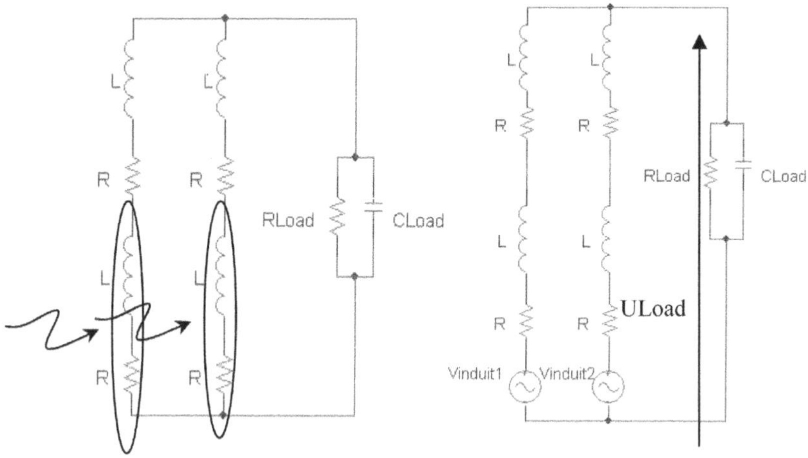

Figure 6.22. *Two antennas of different connections are illuminated*

We again use the theorem of superposition by considering the case or only V_ind_1 is switched on and V_ind_2 is switched off then the case where only $V_{induit2}$ is switched on and $V_{induit1}$ is switched off, and finally during the presence of the two sources of voltage, the result is the sum of two preceding voltages, either:

$$V\_ic = V\_ic_1 + V\_ic_2 = \frac{Rp\_ic}{2Rp\_ic + 2(Rs\_ant + jLs\_ant\omega\_c) \times (1 + jRp\_ic \times Cp\_tot\omega\_c)}(V\_ind_1 + V\_ind_2)$$

Conclusions on the incidences due to voluntary or involuntary coupling of the antennas belonging to the same connection or to neighboring connections.

Pursuant to the superposition theorem, the effects of coupling on different antennas add up provided that this coupling occurs with the same phase (therefore, pay attention to the direction of antenna connection "in strips" because additive or subtractive flow).

6.4.4. *Simplification of the equation*

In the two cases described above, if the resistance of the integrated circuit is as:

$$Rp\_ic >> \left| \frac{1}{jCp\_tot\omega\_c} \right|$$

in the work frequency, the equation therefore becomes:

$$V\_ic = \frac{V\_ind}{2 + 2(Rs\_ant + jLs\_ant\omega\_c) \times (jCp\_tot\omega\_c)}$$

If Q_ant is great (>10) leading to

Ls_ant = Lp_ant.

and by calling:

Lp_ant x Cp_tot = 1/ω_tag^2

it comes to:

$$V\_ic = \frac{V\_ind}{2\left(1 - \omega\_c^2 / \omega\_tag^2\right) + 2jRs\_antCp\_tot\omega\_c}$$

Knowing Cp_tot, we have determined Lp_ant in such a way that the tuning frequency of the tag is equal to the carrier frequency ω_tag = ω_c Therefore:

$$V\_ic = \frac{V\_ind}{2(jRp\_ant \times Cp\_tot \times \omega\_c)}$$

In defining the antenna quality factor at the operating frequency of the mobile telephone by:

$$Q\_ant = \frac{Ls\_ant\omega\_c}{Rs\_ant}$$

As we have tuned the tag (ω_c = ω_tag), we also have:

$$Q\_ant = \frac{1}{Rs\_antCp\_tot\omega\_c}$$

because

$$Ls\_ant \times Cp\_tot \times \omega\_tag^2 = 1$$

Therefore:

$$V\_ic = \frac{-jQ\_antV\_ind}{2}$$

6.4.4.1. *Generalization with a network at "n²" (2, 4, 9, etc.) antennas*

The structure of the antenna network comprises of a total of "n" connections, each having "n" antennas, each individual antenna having the inductance value as "Ls_ant", and of resistance "Rs_ant". Since the superposition theorem is applicable, we will only study the case in which it is supposed that during the presentation of the mobile telephone in the NFC frame/tablet the magnetic coupling is only produced *via* a single antenna and therefore the magnetic flux originating from the mobile induces a d.d.p only in a single antenna of a single connection. That is V_ind is the induced voltage on this antenna.

The general coupling diagram is given in Figure 6.23.

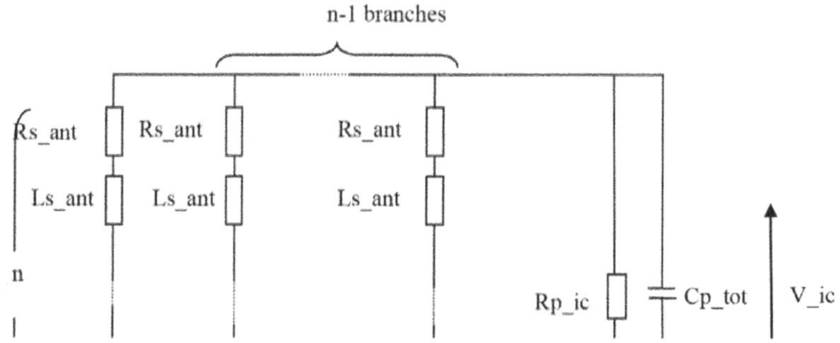

Figure 6.23. *Coupling general diagram*

We now proceed to a series of equivalent diagrams (Figures 6.24–6.26).

1) In each of the n connections, we add the inductances and resistances series.

2) The Thevenin Norton transformation is used for passing from a generator voltage of V_ind to a diagram composed of a J current generator:

$$J = \frac{V\_ind}{nRs\_ant + jLs\_ant\omega\_c}$$

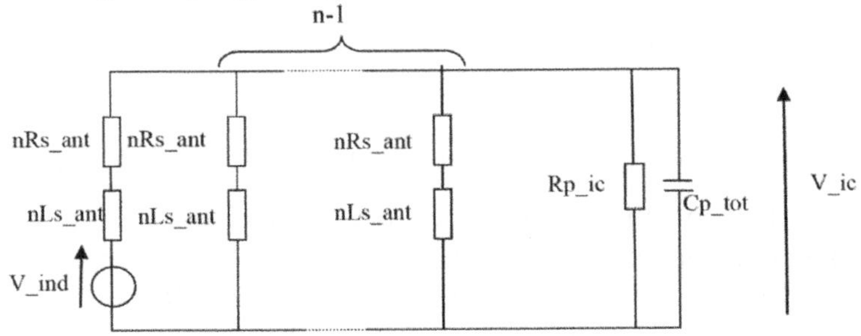

Figure 6.24. *Addition of the n connections*

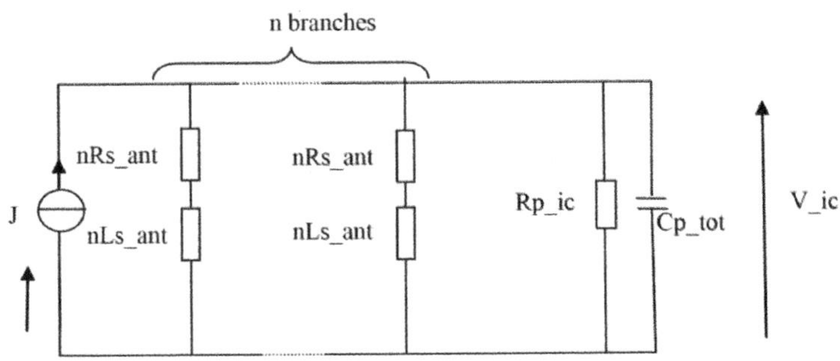

Figure 6.25. *Thévenin–Norton transformation*

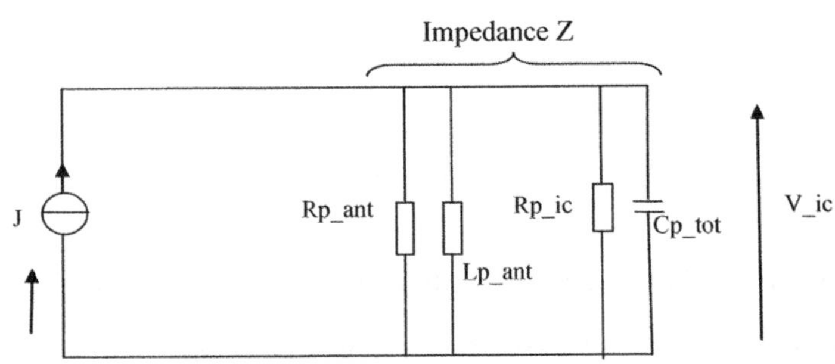

Figure 6.26. *Parallel series transformation*

3) We carry out a last parallel series transformation. The n inductances nLp_ant in parallel give an inductance of Lp_ant. Likewise for the n resistances nRp_ant.

By noting Rp_tot the equivalent resistance in paralleling of Rp_ant and de Rp_ic, we have the Z impedance such as:

$$\frac{1}{Z} = \frac{1}{jLp\_ant\omega\_c} + \frac{1}{Rp\_ant} + \frac{1}{Rp\_ic} + jCp\_tot\omega\_c$$

which is:

$$Z = \frac{jLp\_ant \times \omega\_c}{\left(1 - Lp\_ant \times Cp\_tot \times \omega\_c^2\right) + j\dfrac{Lp\_ant \times \omega\_c}{Rp\_tot}}$$

We now calculate the voltage which appears at the terminals of the integrated circuit:

$$V\_ic = ZJ = \frac{1}{n} \times \frac{1}{\left(1 - Lp\_ant \times Cp\_tot \times \omega\_c^2\right) + j\dfrac{Lp\_ant \times \omega\_c}{Rp\_tot}} \times \frac{jLp\_ant \times \omega\_c}{Rs\_ant + jLs\_ant \times \omega\_c} \times V\_ind$$

Furthermore, the overvoltage coefficient of the tag is Q_tag = Rp_tot/Lp_ant x ω_tag.

In case, the target tuned is on the carrier frequency of the initiator, ω_tag = ω_c, it comes to:

$$V\_ic = \frac{-jQ\_tag}{n} \times \frac{jLp\_ant \times \omega\_c}{Rs\_ant + jLs\_ant \times \omega\_c} \times V\_ind$$

Subject to that the Rs_ant is weak in relation to the term (Ls_ant ω_c), the writing of the equation is simplified into:

$$V\_ic = \frac{1}{n} \text{ x } - j\, Q\_tag \text{ x } V\_ind$$

and in module:

$$\frac{V\_ic}{V\_ind} = \frac{1}{n} Q\_tag = f(n, Q\_tag)$$

OBSERVATION.–

With Lp_ant = 5.3 μH → Lp_ant x ω_c = 451 ohms

Rs_ant = qq ohms << Ls_ant x ω_c = 451 ohms

6.4.4.2. *The sequence*

After this long foretaste explaining the principles, let us now pass to the result of the antenna design:

– first, at the same "L" inductance value of the antenna of the necessary tag for obtaining the good tuning with the internal capacity of the integrated circuit, the total surface value of the S_tag "Ae" tag embracing the magnetic flow received of the initiator;

– then, on the same total surface "Ae" optimize the coil shape in such a way that the couple "N, s" ("N" number of turns and "s" surface of a coil) satisfying the condition of flux above;

– then, seeking to optimize the magnetic coupling value "k" between tag (target) and mobile (initiator) to obtain the greatest operating distance to the same induction supplied by the mobile.

It should be noted that to achieve all that several stages of iterations will be carried out in order to obtain satisfying results.

6.4.4.3. *Estimation of the tag antenna minimum surface, "Ae" to recover the necessary flow to the requirements of the application*

Once again, knowing that the rationalized equation of the magnetic field H(a, r) product by an initiator is equal to:

$$H(a, r) = \frac{1}{[(1 + a^2)^{\wedge}3/2]} \times H(0, r)$$

with:

– "d" distance;

– "r" radius of the initiator antenna;

– "a = d/r" ... d = (a x r).

We can, therefore, study the value of H(a, r) according to the values "r" and "d" or "a" =d/r.

6.4.4.4. *Application in the NFC frame/tablet*

Supposing that the antenna of the initiator mobile is rectangular with a mechanical surface of 50.4 cm^2 and can be reverted to a circular antenna of an equivalent surface (which is not very wrong...) the latter would, therefore, have a radius of r = 4 cm and with *an operating distance of 2 cm* – that would lead to considering a ratio "a" = d/r equal to 2/4 = 0.5, which leads to a value of *H_d/H_0 = 0.716 on the principal axis* ... a case of aligned mobile telephones/tags (cosθ = 1) in the most optimum operating position. This means that in this better position:

1) that in the case H_0 = 1 A/m rms (from the side of the screen), the tag will only hope to have 0.716 A/m functioning at the most;

2) that in the case H_0 = 0.5 A/m (from the rear of the mobile), the tag will only have 0.358 A/m at its disposition to function.

OBSERVATION.– In the case where the mobile and tag plan (of the frame) form an angle of 45° (cos 45° = 0.707), we will only have 70% of these values.

A "glued copy" never did anyone any harm, just "for information" in carrying out the same calculations for a desired operating distance of 3 cm and remaining in the case H_0 = 0.5 A/m (from the rear of the mobile), the tag will only have 0.25 A/m at its disposition to function and it would be necessary to increase the magnetic field emitted by the initiator of 0.358–0.5 A/m (that is a 40% increase) to obtain an unfortunate 1 cm extra operation.

6.4.4.5. *Relations between magnetic fields and inductions and H_th_min*

Knowing that the initiator antenna (the mobile telephone) produces a magnetic field in its center H_d_0 (for example, 0.5 A/m from the rear) and on its main axis, in the air, having the available field value at the desired operating distance "d" (H_d_2cm = 0.358 A/m), we calculate the minimum value of the total surface of a tag *mono-antenne* capable of collecting the flow in order to be able to arrange the max value. V_th_max = 2.27 V rms of the ddp min. necessary for the most reticent operation of the tag integrated circuit.

We calculate the magnetic field of "threshold H_d_th" whose value is given by the formula which has been presented many times:

$$H\_d\_th = \frac{\left[\left(1 - \left(\frac{\omega\_c}{\omega\_t}\right)^2\right)^2 + \left[\frac{1}{Qp2}\right]^2\right]^{1/2}}{\omega\_c \times \mu\_0 \times N2 \times s2} \times (V\_ic\_th)$$

In case the target is tuned on the emitted frequency by the initiator ($\omega\_c = \omega\_t$), the equation becomes:

$$H\_d\_th = \frac{1}{\mu\_0 \times N2 \times s2 \times \omega} \times \frac{1}{Qp2} \times (V\_ic\_th)$$

with	H_0	= 0.5 A/m	rear side of the mobile telephone
that is	H_d_2cm	= 0.358 A/m	
	$Qp2\_min$	= 10.31	tag adjusted in power

$$0.358 = \frac{1}{(4 \times 3.14 \times 10^{-7}) \times (N2 \times s2) \times (85{,}157\ 10^{+6})} \times \frac{1}{10.31} \times (2.27)$$

$$N2 \times s2 = \frac{22.7}{2756.8}$$

$$Ae = N2 \times s2$$
$$= 61.36\ \text{cm}^2 \text{ minimum} \quad \text{.... for the Q2p min}$$

In summary, for the use of a "mono-antenna", the problem that arises is as follows: for the application "NFC frame/tablet"/communication distance 2 cm, it is possible to realize an antenna coil entering the frame mechanics simultaneously:

– its inductance value is L = 5.3 μH;

– its unloaded quality coefficient Q_ant_min = 20.62;

– its total min. surface that is Ae = 61.36 cm².

By using:

– the mobile telephone producing a magnetic field on its back H_0 = 0.5 A/m, that is H_2cm = 0.358 A/m;

– for an NFC Forum tag of T1T type (for example, TOPAZ integrated circuit of innovision) whose value of V_ic_th_max is 2.27 V eff.

6.4.4.5.1. Observations regarding a multi-antenna network application

We have shown that in the case of a multi-antenna network of type "n² antennas":

$$\frac{V\_ic}{V\_ind} = \frac{-jQ\_tag}{n}$$

Yet, V_ind the sinusoidal voltage induced on an antenna is as already indicated above:

$$V\_ind = -j\,\omega\,B\_d \times S\_ant = -j\,\omega\,\mu_0 H\_d \times S\_ant$$

where S_ant is the surface of the illuminated tag antenna by mobile telephone (N_ant x s_ant) and B_d is the magnetic induction in the distance d of the telephone antenna. The magnetic field radiated by the mobile telephone is, therefore, in an absolute value:

$$H\_d = \frac{V\_ind}{\omega\mu_0 S\_ant} = \frac{nV\_ic}{Q\_tag \times \omega\mu_0 S\_ant}$$

which leads to, in relation to a mono-antenna system, the appearance of the multiplication factor "n" in the equation of the surface necessary for good operation:

$$S\_ant = n \times \frac{V\_ic}{Q\_tag \times \omega\mu_0 H\_d}$$

To conclude this topic, it is good to know that in order to conserve a reasonable antenna surface the appearance of the factor "n" can be totally or partially compensated by a simultaneous increase and of the same size of the tag quality coefficient value Q_tag, the theoretical max. Q_tag being the only value linked to the value R_ic.

6.4.4.5.2. Optimization of the couple N_ant and s_ant

In order to proceed to the optimization of the couple N_ant, s_ant, the maximum mechanical sizes of the network base antenna will need to be defined. For this purpose, we will define several parameters.

6.4.4.5.3. Choice of the number of antennas and incidences of this number

Incidence of the number of antennas

The larger the number of antennas, the better the distribution of the magnetic flow reception originating from the mobile.

Mechanical distribution/disposition of the antennas

Given that the principle "parallel series" adopted for the realization of the antenna network needs to have an n^2 number of antennas, we can only have solutions of type 4, 9, 16, 25, 36, etc., antennas.

4 antennas solution

Several possibilities can be envisaged:

– one per corner;

– one per frame/tablet side.

In one case or another, the cover of the space cannot be complete … but one solution or the other could be envisaged to make small size frames/tablets.

9 antennas solution

This antenna distribution choice permits us to offer a larger space cover for communication to the initiator. Of course, an odd number is not divisible by four and the principal issue is to define the geographical distribution on the frame/tablet.

A solution of $n^2 = 9$ is found, for example, in:

– 3 on the large superior side;

– 2 on each of the small sides;

– 2 on the other large side (this is placed on the desk) which leaves a little space for arranging the integrated circuit at the center and carrying out the antenna connections in the integrated circuit.

Mechanical shape of the basic antenna

Squared-rectangular

Due to the shape of the frame or tablet, the rectangular solution is preferred in order to obtain a better coupling because of the shape of the antennas presented in the mobiles. Although having a horizontal or vertical orientation, this will not always be optimal at the magnetic coupling level between initiator and tag.

The maximum limit of antenna width is given:

– by the max width from the frame/tablet edge;

– by the passage of certain tracks if the printed circuit is not two-sided.

Shape versus coupling

In relation to the mobile antenna, we will estimate the couplings between base station and tag depending on their respective antennas a little farther.

ATTENTION.– We ask you to pay attention in relation to mobiles of other models and brands whose antennas can be very different in shape and performance.

6.4.4.5.4. Arrangement of the 9 antennas

The arrangement is shown in Figure 6.27.

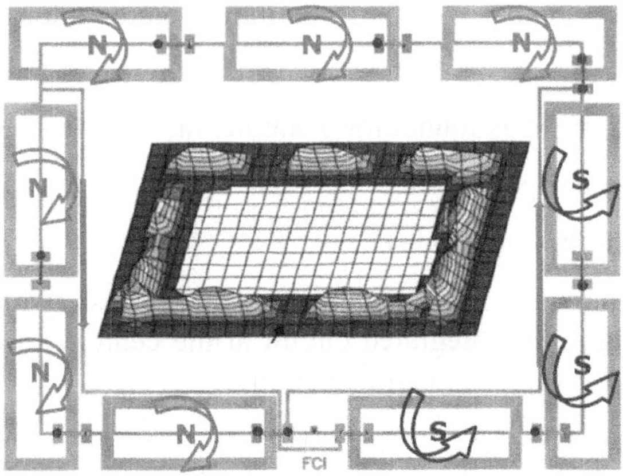

Figure 6.27. *9 antennas solution*

For all the reasons given above, the rectangular antenna is the one that is adjusted the most, it is that which will facilitate the best realization of the minimum useful surface Ae. In our example, having taken into account the internal and external mechanical sides of the frame or tablet, the maximum sizes of the target antenna will not be able to exceed:

– 25 mm wide (because of the width of the frame/tablet edge);

– 44 mm long (if we want to arrange two antennas on the small side and three antennas on the large side of the frame).

That is a surface value of the *average coil of 23 mm x 40 mm = 9.2 cm²* and considering that the required minimum surface for the application Ae_min is 61.36 cm² that leads to a number of minimum turns of:

N2 = 61.36/9.2
= 6.67 min
= 7 turns min

We are now going to examine if that is realizable and, if yes, how.

When applying the classic formulas, see Chapter 7 for rectangular antennas and after having "used" a specific optimization software, Table 6.9 summarizes a first flow of "coherent" results of coil mechanical sizes, inductance value and minimum surface in order to carry out a design in view of industrialization.

Threshold Res. Frequency	13.56	MHz
Chip capacitance (Threshold)	21.1	pF
Connection Capacitance	2	pF
Coil capacitance	2.9	pF
Parallel capacitance	26	pF
Overall length	40	mm
Overall width	23	mm
Track thickness	35	µm
Track width	200	µm
Gap between tracks	200	µm
Number of turns	9	
Turn exponent	1.8	
Average coil area	706.16	mm²
L	**5.37**	µH

Table 6.9. *Example of antenna winding*

OBSERVATION.– Due to the mechanical sizes "length x width" of the antenna, to succeed in simultaneously obtaining the value couple of the inductance and surface Ae min, it is necessary to arrange nine coils instead of 7 min.

6.4.4.5.5. Some observations on the table of content

Antenna ohmic resistance:

Continuous resistance. We can calculate the continuous resistance value (DC) of the antenna:

R_dc = (ρ N_ant l)/s

The resistivity of the copper Cu and aluminum Al is, respectively:

Cu = 1.8 10^-8 et Al = 3 10^-8 ohm meter

Length of the antenna conductor winding:

– L average of a coil = (3.6 cm + 2 cm) x 2

– L total = 11.2 cm x 9 = 1 m

Section of the antenna:

$$200\ \mu m \times 35\ \mu m = 7000\ 10^{\wedge}{-12}\ m^2$$

For example, for a realization in aluminum:

$$R\_ant\_dc\_Al = 3\ 10^{\wedge}{-8} \times (1/(7000\ 10^{\wedge}{-12})$$
$$= \text{about } 4.2 \text{ ohms}$$

This value of R_ant_dc_Al, lower in power than that necessary to the adjustment of the tag, facilitates the recovery of the value Q_ant and, induced in the same flux, to present a little more voltage in the integrated circuit. For this reason, compensate in part the division by 1/n due to number of connections. In fact with R_ant = 4.2 Ω, Q_tag = (5.3 10^-6 x 85.157 10^6/(4.2 + 27)) = 14.47 instead of 10.31 either 1.4 or a compensation of about the half in relation to the increase of n = 3

Average surface

In the above example, due to nine coils, the total surface of the tag antenna is:

s_moy x 9 = 7.06 (see Table 6.9) x 9
= 63.5 cm^2 (to be compared with 61.36 cm^2 theoretically)

It is the best surface value that we can obtain by conserving the inductance optimal value, but other compromises exist and they are proposed below.

6.4.4.5.6. Choice of the winding technology

For reasons of cost, it is necessary to study the choice of winding technology between aluminum and copper.

Technology	"Aluminum"	"Copper"
Thickness of track	30 μm min	35 μm min
Width of track	250 μm min	150 μm min
Gap between track	250 μm min	150 μm min
Width of the tag (target) antenna: maximum 23 mm to enter in a mechanical size of 27 mm and minimum 20 mm to be able to pass tracks of connections without using a two-sided circuit.		

Table 6.10. *Incidences of the winding technology*

Taking into account these constraints, Tables 6.11 and 6.12 give overviews of the calculations seeking to maximize the antenna total surface at the same inductance value.

Aluminum						
Max width	mm	**23**	**23**	**20**	**20**	**20**
Max length	mm	43	54	39	47	59
Number of turns		9	8	10	9	8
Inductance	µH	**5.27**	**5.28**	**5.25**	**5.25**	**5.31**
Surface Average	cm²	7.12	9.50	5.10	6.58	8.80
Surface Total	cm²	64.08	**76**	51	59.2	**70.4**
Resistance	Ω	4	4.4			~ 4
Coefficient Quality						
Conclusions		True	**Yes**	No	No	**Yes**

Table 6.11. *Overview of aluminum winding*

Copper						
Max width	mm	**23**	**23**	**21**	**21**	**21**
Max length	mm	35	45	30	37	47
Number of turns		9	8	10	9	8
Inductance	µH	**5.31**	**5.29**	**5.32**	**5.26**	**5.31**
Surface Average	cm²	6.55	8.77	4.86	6.27	8.29
Surface Total	cm²	58.95	**70.16**	48.6	56.43	**66.32**
Resistance	Ω	3.5	4			~ 3.5
Coefficient Quality						
Conclusions		No	**Yes**	No	No	**True**

Table 6.12. *Overview of copper winding*

The ohmic and coefficient values of quality of all these antennas are close enough.

Furthermore, we must not forget to consider the ohmic resistance value of the linking wires of the antenna connections, the farthest mechanically from the integrated circuit. Indeed, the wire length of an antenna is about 1–1.10 m and the links can reach up to 2 × 25 cm = 0.5 m. Therefore, if possible, increase the width of the track to return to this antenna connection toward the integrated circuit.

6.4.4.5.7. In conclusion

With the aim of guaranteeing the operating distance, we quantify the couplings obtained between tag antenna surfaces and those of the mobiles for an orientation of the determined mobile.

As indicated, the coil winding of the mobile antenna (initiator) is rectangular in shape 4.5 cm × 11.2 cm, that is a surface per coil Ae of about 50.4 cm². By ease of calculation, if this antenna (in the air) was circular, its radius "r1" would have been about 4 cm and the magnetic field value in its center would have been:

$$H\_0\_cm = (N1 \text{ x } I\_L1s)/2 \text{ r1}$$

Knowing the number of antenna turns of the mobile (N1 = 3 turns); it is easy to estimate the value of the current "I_L1s" which would have permitted the field to produce (supposing the worst):

$$H\_0\_\text{screen side} = 075 \text{ A/m} \rightarrow 0.75 = (3 \text{ x } I\_L1s)/(2 \text{ x } 0.04)$$

that is a current:

$$I\_L1s = 20 \text{ mA rms}$$

Furthermore, if r1 = 4 cm and N1 = 3, the inductance value of this circular antenna (on air) would be L1 given by the classic formula:

$$L1 = 2\,(1)\,[\,\ln(l/D) - 1.07\,] \text{ x } (N^2)$$

with			
with	N	= number of rotations	= 3
	Ae	= section of a turn	= 50.4 cm²
	l	= 2 x pi x r1	= 25.12 cm
	D	= diameter of wire	= 200 µ

$$L1\,(nH) = 2 \text{ x } (25.12) \text{ x } [\,\ln(25.12/0.02) - 1.07\,] \text{ x } (3^2)$$

that is about L1 = 2.742 µH. To simplify the calculations, we undertake for L1 = 2.5 µH.

The d.d.p. at the terminals of the antenna of the mobile telephone (initiator) would be equal to:

$$V1 = (L1s \times \omega) \times I1 = (2.5\ 10^{-6} \times 85.156\ 10^{6}) \times 20\ 10^{-3} = 5.53 \text{ V eff}$$

Knowing that the d.d.p received in the air to the terminals of the antenna of tag V20_typ must be equal to:

$$V20\_typ = V\_ic\_typ/Q\_tag = 2.1/10 = 0.21 \text{ V rms}$$

We can, therefore, calculate the order of magnitude of the k_typ coupling which is necessary to the application:

$$V20\_typ = V1 \times k\_typ \times \sqrt{(L2/L1)}$$

$$k\_typ = (0.21)/(5.53 \times \sqrt{(5.3/2.5)}) = 2.6\ \%$$

In this case, the mutual value would, therefore, be worth:

$$M\_min = k \times \sqrt{(L2 \times L1)} = 0.026 \times 10^{-6} \times \sqrt{(5.3 \times 2.5)} = 106 \text{ nH}$$

6.4.4.6. *Value of the operating typical minimum distance*

The value "k" permits the estimation of the maximum theory distance of operation possible (see Chapter 1 on the subject):

$$k\_min\_typ = \left(\mu\_0 \times \frac{r^2}{(r1^2 + d\_max^2)^{3/2}} \right) \times N1 \times N2 \times s2 \times \left(\frac{1}{\text{racine de } L1 \times L2} \right)$$

with:

k_min_typ	= 2.6 %	$\mu = \mu\_0 = 4 \times \pi \times 10^{-7}$	
N1	= 3	L1 = 2.5 µH	r1 = 4 cm
N2	= 9	L2 = 5.3 µH	s2 = 50.4 10^{-4} m^2

we obtain:

$$(r1^2 + d\_max^2)^{3/2} = \left(\frac{2.73466}{950}\right) = 28.8\ 10^{-4}$$

$$(r1^2 + d\_max^2) = 202.4 \times 10^{-4}$$

$$d\_max^2 = (202.4 - 16)\ 10^{-4}$$

$$\text{soit} \quad d\_max = 13.6 \text{ cm}$$

The above estimation of this distance does not include the tolerances of the components and does not, therefore, indicate the value of the maximum distance "minimum" guarantee. For this purpose, it is necessary to calculate all the partial derivatives of the equation k_min above:

– first, it is supposed that N1, N2 do not have tolerance (N1 = 3 turns and N2 = x turns will be realized in printed circuit);

– the same goes for s2, same reason, same punishment;

– therefore, the only thing remaining is to take interest in value tolerances of L1 and L2. By being large, generally these tolerances are 10% for each of them. The squared root of their product is, therefore, also 10%.

Obviously, we must also integrate the tolerances of the value of current. It circulates in the antenna of the base station, the variations of the supply voltage which supplies the latter, the most unfavorable value of V_ic, etc., in brief all this facilitates the estimation of the "*worst case*".

7

The Initiator–Target Couple and its Couplings

Let us quickly recap some essential points of near-field communication (NFC) applications – in particular, that:

– the associated antenna of an "NFC device" (initiator or target) may have absolutely any mechanical shape; it can be large, small, round, square, crooked, etc.; and therefore may have absolutely any form factor. It can be a paper target, a TV set, a PC, a photo frame, a tablet, a car dashboard, a keyring, a pendant, an NFC bracelet and, of course, a mobile phone;

– the values of the magnetic fields may evolve during the application;

– the distance, and therefore the magnetic coupling, may evolve during the application;

– the NFC device's immediate/direct environment may be any we choose, and particularly *hostile to wave propagation*;

– the antenna of an NFC device (an initiator or target) may be either *deliberately not tuned* to the carrier frequency of 13.56 MHz, or at any time may be *accidentally detuned* by the environment;

– NFC devices are sometimes arranged in stacks. Such may be the case, for example, with multiple mobile phones (one personal and one for work) in a bag as it passes in front of a public transport validator. Indeed, on the one hand, we need to manage electrical incidents (couplings) between NFC device antennas, and on the other hand, manage the different radio frequency (RF) collisions of initiators, of non-uniqueness of the identifiers of NFC devices, of data transmitted by each one, and finally, manage the problems of polling of protocols and applications;

– finally, for an "NFC device", the onboard application can engender any kind of *function per slice of time* (first working as an initiator and then a simple target and then card emulation for the next 5 min, for example, etc.) ... and the function of the antenna must obey.

In the wake of this brief, matter-of-fact recap – which is all too often overlooked by numerous system designers, and gives rise to many applicational – and interoperability problems – let us return to the physical and electronic fundaments of the history of the lower layers of NFC and to the key points from the above sections.

Here, then, are the five main topics of this chapter, and their direct relations with the design of an antenna which we thought was so simple:

– form factor;

– deliberate tuning and/or deliberate or involuntary detuning ("stacks");

– an environment hostile to wave propagation;

– different functions for each slice of time;

– variation of distance, and therefore of the coupling.

7.1. Circuits and their couplings

Let us now dedicate the following sections to circuits (for the sake of simplicity, we will start with two: the initiator and the target), and the ways in which they can be coupled electromagnetically in NFC applications:

– *random shapes/geometries of circuits*: luckily, circuits are usually flat and of simple shapes – circular, square, etc.;

– *relative positions of circuits in space*: again, luckily, their spatial positions are generally considered as being situated in parallel planes, and (very) often, the primary axes of the circuits are (supposed to be) essentially coaxial ... except in the case of mobile phones, because we do not know where the target antennas are;

– *distances between circuits*: although they are considered to be located on parallel planes, and the axes are (very) often coaxial, the distance "d" between the coupled initiator and target circuits is a parameter that is either constant or variable during the time when the target is presented to the initiator … with all the attendant consequences;

– *tuning of the circuits*: the (two) coupled circuits represented by the initiator–target couple are generally tuned. Their respective tunings can be performed indifferently on "far-removed" frequencies (e.g. ~14 and ~30 MHz), "near" frequency (e.g. 13.56 and ~16 MHz), "very near" frequencies (e.g. ~13.56 and ~14.5 MHz) or indeed "identical" (e.g. 13.56 and 13.56 MHz).

In summary, see Figure 7.1.

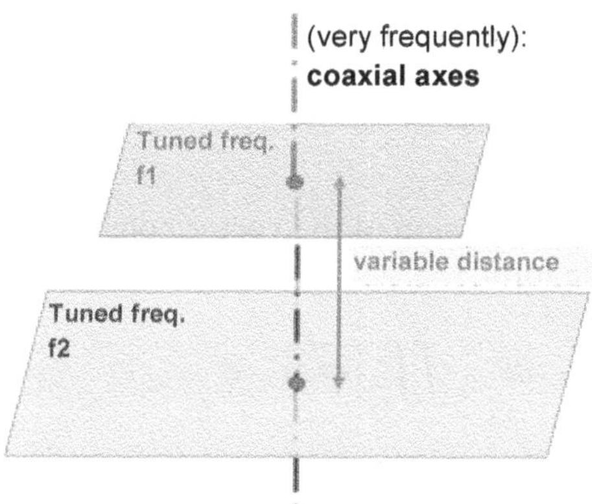

Figure 7.1. *Stack of couplings*

Nevertheless, let us begin with a few very useful fundamental recaps.

7.1.1. *Mutual induction and mutual inductance*

The phenomenon of *mutual induction* characterizes the influence of one *magnetic* circuit on another and expresses the fact that a variation of current in a magnetic circuit can lead to the appearance of a voltage in another magnetic circuit. By definition, mutual induction between two circuits is defined by the ratio between the *flux* created by an *electric dipole* passing through a second dipole and the current having created that flux (Figure 7.2).

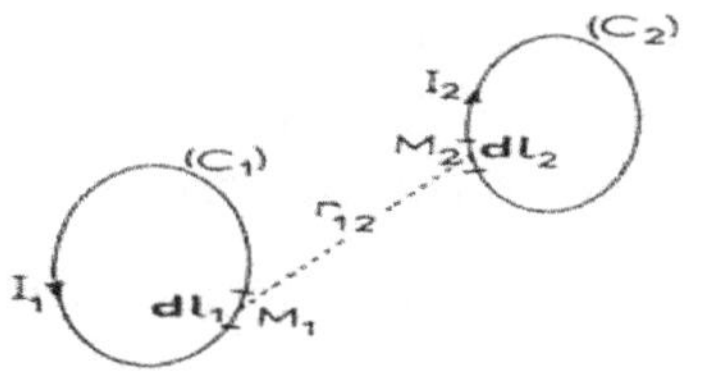

Figure 7.2. *Principle of mutual induction*

To put this into figures, consider two circuits (loops of antennas) C1 and C2 traversed by currents I1 and I2. The current I1 circulating in C1, throughout the whole space, produces a magnetic field H1 and a magnetic induction B1, and the flux generated by the induction B1 through C2 is written as Φ12. Its value is:

$$\phi_{12} = \iint_{s_2} \vec{B_1} \cdot \vec{dS}$$

Using the classic expression of the vector potential $\vec{A}$, we can rewrite the above relation in the form:

$$\phi_{12} = \iint_{s_2} \vec{B_1} \cdot \vec{dS} = \iint_{s_2} \overrightarrow{rot\vec{A_1}} \cdot \vec{dS}$$

Similarly, using Stokes' theorem, the above relation becomes:

$$\phi_{12} = \iint_{s_2} \overrightarrow{rot\vec{A_1}} \cdot \vec{dS} = \oint_{c_2} \vec{\mathrm{A}_1} \cdot \vec{dl_2}$$

In addition, the expression of the vector potential A1 generated by a linear circuit C1 traversed by a current I1 is written as:

$$\vec{A_1} = \frac{\mu_0}{4\pi} \oint_{C_1} I_1 \frac{\vec{dl}}{r_{12}}$$

Thus, the (true) final expression of the flux Φ12 is the following curvilinear double integral on C1 and C2 (dl_1 and dl_2 being the elements of length of the curves C1 and C2):

$$\phi_{12} = I_1 \cdot \left(\frac{\mu_0}{4\pi} \oint_{c2} \oint_{c1} \frac{\vec{dl_1} \cdot \vec{dl_2}}{r_{12}} \right)$$

which is written as: Φ12 = L_{12} I1… and the value of the mutual induction *L12 – which, later on, will be written as Mutual "M"* – between the circuit C1 and the circuit C2 is therefore equal to Φ12/I1 = L_{12}:

$$L_{12} = \frac{\mu_0}{4\pi} \oint_{c2} \oint_{c1} \frac{\vec{dl}_1 \cdot \vec{dl}_2}{r_{12}} = M$$

EXAMPLE.–

$$\begin{aligned} Area &= \int_0^R \int_0^{2\pi} r \; dr \; d\theta \\ &= 2\pi . R^2/2 \\ &= \pi R^2 . \end{aligned}$$

Note that M = L12 is unchanged by permutation of the indices 1 and 2 in the calculations, so:

Φ21 = L21 I2 = L12 I2

The parameter "M" represents the "mutual" influence (mutual incidence of a primary current I1 or a primary voltage V1 on a secondary voltage V2, and *vice versa*). As just stated, the action of this mutual can be assimilated to the *equivalent of the physical presence of an inductance* "M" between the primary and secondary electrical circuits. Thus, its unit is the "Henry".

If the medium's magnetic permeability μ_r is homogeneous, the magnetic fluxes are proportional to the currents i1 and i2, and the proportionality coefficients involved are:

– the primary inductance L1;

– the inductance of the secondary L2;

– the mutual inductance M between the two circuits.

7.1.2. *Perfect mutual*

If the whole of the flux produced by the primary inductance L1 runs through the whole of the secondary inductance L2, then the mutual – M – is said to be perfect. If we adopt this hypothesis, it is easy to quantify this magnetic mutual M with non-tuned coupled circuits:

– the circulation of I1 induces an emf = - j M ω I1 in the secondary;

– the circulation of I2 induces an emf = - j M ω I2 in the primary.

In this case, by applying a voltage V1 to the primary, the laws of networks applied to the primary and secondary circuits enable us to write:

– at the primary V1 - j M ω I2 = I1 (R1 + j X1) where X1 = L1 ω;

– at the secondary - j M ω I1 = I2 (R2 + j X2) and X2 = L2 ω.

By drawing I2 from the equation of the secondary and substituting its value back into the equation of the primary, after expansion, we obtain:

$$V1 = \left[I1 \left(R1 + R2 \frac{M^2 \omega^2}{R2^2 + X2^2}\right] + j \left[\left(X1 - X2 \frac{M^2 \omega^2}{R2^2 + X2^2}\right)\right]$$

In order for this equation with complex variables to be valid, its imaginary part must be equal to zero, meaning that:

$$X1 = X2 \left(\frac{M^2 \omega^2}{R2^2 + X2^2}\right)$$

If R2 is smaller than L2 ω (which is practically always the case, for us), then $R2^2$ is negligible in comparison to $X2^2$, and we obtain:

$$L1\ \omega = \frac{M^2 \omega^2}{L2\ \omega}$$

Thus, in conclusion:

$$M = \sqrt{(L1\ L2)}$$

7.1.3. *Non-perfect mutual*

Let us now examine the most usual case (which we will normally deal with in the application field of NFC), where only a portion (large or small) of the flux produced by the initiator flows through the target. In this case, the total flux (Φ_total) produced by the initiator's coil L1, traversed by the current I1, is divided into two, between a useable flux (Φ_useable) traversing L2 of the target, and a leakage flux (Φ_leak).

The fluxes produced by the currents i1 and i2 are given in Table 7.1.

	Flux at the primary	Flux at the secondary
Only the primary current i1 circulates	Φ11 = L1 i1	Φ12 = M i1
Only the secondary current i2 circulates	Φ21 = M i2	Φ22 = L2 i2
Both currents i1 and i2 circulate	Φ1 tot = Φ11 + Φ21	Φ2 tot = Φ12 + Φ22

Table 7.1. *Fluxes produced by currents i1 and i2*

7.1.3.1. *Leakage flux*

Let us distinguish two parts in the flux of the primary Φ11 – one Φ'11 which corresponds to the field lines common to the two circuits, and the other f1 corresponds to the field lines passing through the primary alone rather than passing through the secondary (Figure 7.3).

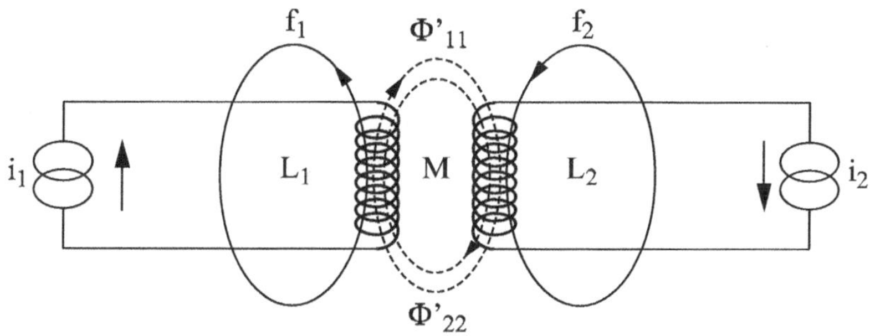

Figure 7.3. *Leakage flux*

The same distinction can be drawn in terms of the flux Φ22 with two components: Φ'22 and f2. In conclusion, only the parts Φ'11 and Φ'22 of the initial fluxes Φ11 and Φ22 correspond to the coupling, while f1 and f2 are leakage fluxes. This phenomenon can be quantified by introducing a certain parameter known as the "coupling index, n" characterizing the "coupling" between L1 and L2. As we will see later on, this coupling, which is representative of the whole assembly of components, could be said to be "tight", "critical" or indeed "weak", and obviously its value depends on the shapes, material, distances, magnetic environments, etc. By setting:

– N1 = number of turns of the initiator antenna;

– N2 = number of turns of the target antenna;

– s2 = surface of a turn of the target antenna;

– B_d = induction at the distance "d" between the coils.

Considering the magnetic flux, we can write that, for the new value of the mutual M':

$$M' \times I1 = N2 \times \Phi 21 = N2 \times (B\_d \times s2)$$

7.1.3.2. *Examples: coils of distant coaxial circular antennas*

For example, in the case of coils of coaxial circular antennas, the value of B_d created by a coil L1 in the initiator with radius r in a coil L2 in the target at a distance d, at its center and on the axis, is known (see below) and enables us to calculate M'.

$$M' = N2 \times [(\mu \times \frac{r^2}{2\,(r^2 + d^2)\text{^}3/2} \times N1)] \times s2 \Longrightarrow M' = f(r, d)$$

7.1.3.2.1. Fortunately, there is often an easier way

Sometimes, to estimate M, with a little bit of luck, instead of having to calculate the lengthy and difficult curvilinear integrals to calcuate mutuals such as those presented a little earlier, such as:

$$L_{12} = \frac{\mu_0}{4\pi} \oint_{c2} \oint_{c1} \frac{\vec{dl_1} \cdot \vec{dl_2}}{r_{12}}$$

it is easier to directly encode the magnetic fluxes which are involved. For example (Figure 7.4), when the L1 (initiator) and L2 (target) (respectively, with N1 and N2 turns) are mechanically basically circular, with radii r1 and r2, coaxial and, in view of the distance separating them, which is nearly concentric (this is exactly the case when we measure the field at the center of an antenna explained in Chapter 8), this simplifies the calculations.

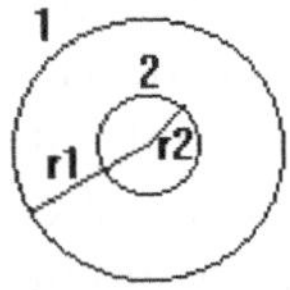

Figure 7.4. *Encoding of magnetic fluxes*

Indeed, we can write (more) easily that (theoretically) the magnetic field H1 produced by C1, L1 (circular initiator) at its center is then:

$$H1 = \frac{N1\ i}{2\ r1}$$

and thus the magnetic induction will, therefore, be B1 = μ H1, and hence the flux generated by B1 through C2 (target with surface area S2 = N2 π r2²), written as Φ12, will be:

$$\Phi 12 = B1 \times S2$$

$$\Phi 12 = \mu \left(\frac{N1\ i}{2\ r1}\right) (N2\ \pi\ r2^2)$$

$$\Phi 12 = \frac{\mu\ \pi\ N1\ N2\ r2^2}{2\ r1}\ i$$

Thus, finally, the mutual M = (Φ12/i) between two coils that are, mechanically, essentially circular, coaxial, and hence, the practically concentric distance separating them is equal to:

$$M' = \frac{\mu\ \pi\ N1\ N2\ r2^2}{2\ r1}$$

EXAMPLE.–

– initiator coil B1: N1 = 5.5 r1= 1.65/2 = 0.825 cm

– target coil B2: N2 = 5 r2 = 0.5 cm

$$M' = \frac{\mu\ \pi \times 5.5 \times 5 \times (0.5^2)\ 10\text{^-2}}{2\ (0.825\ )}$$

$$M' = 164\text{nH}$$

NOTE.– Obviously, if the coils are not strictly circular and/or of particular shapes (square, rectangular, etc.), the (longer) calculation by curvilinear integrals is needed to determine the literal expression of the mutual.

7.1.4. *Coupling coefficient "k"*

In the (very common place) case where the mutual M' is not perfect and has a lower value than that determined in the previous section – in fact *x*% of its maximum – we characterize the quality of the coupling of systems using circular antennas arranged in totally arbitrary positions by introducing a correction factor known as the coupling coefficient "k", defined below.

$$k = \frac{M'}{M} = \frac{\sqrt{(\Phi 12 \,.\, \Phi 21)}}{\sqrt{(\Phi 11 \,.\, \Phi 22)}} = \frac{M'}{\sqrt{(L1 \times L2)}} \quad \text{where k is in \%}$$

Its rigorous expression is given by Neumann's general formula (it is the same as above, but richer and more elegant):

$$k = \frac{\oint_{\Gamma_1}\oint_{\Gamma_2} \frac{\vec{dl_1}\,\vec{dl_2}}{r_{12}}}{\sqrt{\oiint_{\Gamma_1} \frac{d^2\vec{l_1}}{r_1} \oiint_{\Gamma_2} \frac{d^2\vec{l_2}}{r_2}}}$$

where:

– dl_1 and dl_2 length elements of each antenna;

– r_{12} distance between the antenna loops;

– r_1 and r_2 the respective radii.

NOTE.– By principle, we always have "k" less than 1, because indeed, k equal to 1 is obviously an ideal case which is impossible to achieve, because there is always leakage.

The coupling coefficient "k", therefore, is a parameter which depends only on mechanical parameters. Indeed, the values of M, L1 and L2 (the inductances) are only linked to considerations of shape, dimensions, distance, number of turns, etc.

For example, *after careful calculations*, in the case of two colinear circular antennas, where:

– "r" (*equivalent*) is the radius of the initiator antenna, (*a square with side length 19 cm is equivalent to a circle with radius ~10.5 cm*);

– "d" is the distance between the centers of the two coils of the initiator and target antennas;

– "s2" is the value of the surface of a turn of the target antenna;

– L1o and L2o are the values of the inductances per turn – respectively, of L1 and L2;

– L1o = L1/N1 and L2o = L2/N2.

$$k = [(\mu \times \frac{r^2}{2\,(r^2 + d^2)^{3/2}}) \times s2] \times \sqrt{(\frac{1}{L1o \times L2o})} = g(r, d)$$

It should also be noted that when the target is placed at the center of the coil of the initiator antenna (d = 0), we obtain the maximum coupling coefficient which is equal to (often a few %):

$$k\_o = [(\mu \times \frac{1}{2\,r}) \times s2] \times \sqrt{(\frac{1}{L1o \times L2o})} = g(r, d = 0)$$

7.1.4.1. *Estimation and measurement of the value of k*

Using the above equations, it is easy to calculate the value of "M'", and therefore of "k", when we know the mechanical data (distance, etc.), but this proves much more difficult when we decide to introduce, into the equations, the magnetic parameters of a medium (metal parts, magnetic screens, etc.) in which the setup is immersed. Thus, with a view to avoiding these latter setbacks, below we show a simple way to obtain a very good estimation of these values using a measure which is very easy to implement.

With this in mind, let us look at the case of non-tuned coupled circuits where the secondary is not charged (no current I2 and therefore no physical repercussion of the circulation of the current I2 on the primary, so V2 = V20). Thus, we can write:

$$z1 = R1 + j\ L1\ \omega$$
$$i1 = V1/z1$$
$$v20 = -j\,M'\ \omega\ I1$$

By taking the moduli of the complex values, we obtain:

$$V20 = M' \times \omega \times I1$$
$$V20 = [\,k \times \sqrt{(L1 \times L2)}\,] \times \omega \times I1$$

and, knowing that $I1 = V1/Z1$:

Thus:

$$I1 = V1/[\sqrt{(R1^2 + (L1 \times \omega)^2)}]$$

Supposing that $Q1 = (L1\ \ \omega)/R1$ is greater than 10 (which is always the case in NFC applications), meaning that $R1^2 \ll (L1 \times \omega)^2$, we obtain:

$$I1 = V1/(L1 \times \omega)$$

Thus:

$$V20 = [[\,k \times \sqrt{(L1 \times L2)}] \times \omega\,] \times [\,V1/(L1 \times \omega)]$$

Hence:

$$k = \frac{|V20|}{|V1|} \times \frac{\sqrt{L1}}{\sqrt{L2}}$$

measured in a place in the medium – an equation in which V1, V20, L1, L2 are the parameters which are easy to measure in the applications (see example a little later on).

7.2. Tuned circuits coupled by mutual induction

The general electrical diagram pertaining to two non-identical coupled resonating circuits (e.g. "initiator + target") is (almost) as follows (Figure 7.5).

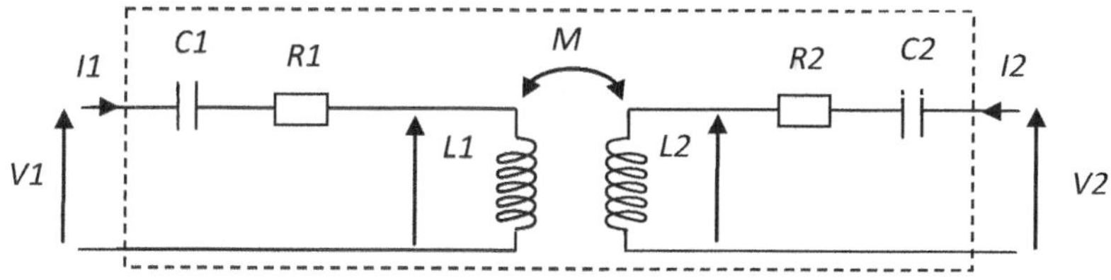

Figure 7.5. *Tuned circuits coupled by mutual induction*

7.2.1. *Why "almost"?*

First, let us consider the coupled system in the form of a non-loaded quadripole (i.e. without taking account of the final load constituted by the target's integrated circuit... meaning just before the threshold field H_d_th of the target is reached – the threshold voltage of the circuit V2o). Now, let us use the following notations:

$$x = 1/L1\ C1\ \omega^2 \qquad \text{and} \qquad Q1 = L1\ \omega/R1$$
$$y = 1/L2\ C2\ \omega^2 \qquad\qquad\qquad Q2 = L2\ \omega/R2$$

When we express the two loops of the quadripole in equation form, we obtain:

$$\begin{aligned}
V1 &= I1\ (R1 + j\ X1) + j\ M\omega\ I2 \\
-j\ M\omega\ I1 &= I2\ (R2 + j\ X2) \\
V1 &= (R1 + j\ (L1\omega - 1/C1\omega)\ I1 \quad + j\ M\ \omega\ I2 \\
&= j\omega\ [L1\ (1 - x + 1/jQ1)\ I1 + M\ I2] \\
V2 &= j\ M\omega\ I1 + (j\ L2\omega + 1/jC2\omega + R)\ I2 \\
&= j\ \omega\ \ [M\ I1 + L2\ (1 - y + 1/jQ2)\ I2]
\end{aligned}$$

which gives us the following matricial formula:

$$\begin{matrix} V1 \\ \\ V2 \end{matrix} = j\omega \begin{vmatrix} L1\ (1 - x + 1/jQ1) & M \\ & \\ M & L2\ (1 - y + 1/jQ2) \end{vmatrix} \begin{matrix} I1 \\ \\ I2 \end{matrix}$$

This matrix has the determinant "Det":

$$Det = j\ \omega\ \ [L1(1 - x + j/Q1)\ L2(1 - y + j/Q2) - M^2]$$

$$Det = j\ \omega\ L1\ L2\ [(1 - x + j/Q1)(1 - y + j/Q2) - M^2/L1L2)$$

and including the coupling coefficient $k = M/\sqrt{L1L2}$

$$Det = j\,\omega\, L1\, L2\, [(1 - x + j/Q1)(1 - y + j/Q2) - k^2]$$

It is fairly easy to demonstrate that this determinant never takes the value of 0, and that the matrix can always be inverted.

7.2.2. *Coupling index "n"*

In light of relation $M = k\sqrt{L1L2}$, it is tempting to define a new relation linked to the coupling – known as "n", the coupling index – by the formula:

$$n^2 = \frac{M^2}{L1(1 - x + j/Q1)\, L2(1 - y + j/Q2)}$$

$$n^2 = k^2 \frac{Q1}{(Q1(1 - x) + j)} \times \frac{Q2}{(Q2(1 - y) + j)} \quad \text{complex value}$$

Thus, $n^2 = k^2\,(A1\, A2)$ with

$$A1 = \frac{Q1}{(Q1(1 - x) + j)} \quad \text{and } A2 = \frac{Q2}{(Q2(1 - y) + j)}$$

The original (physical) coupling coefficient "k" is, therefore, multiplied by an overall gain A = A1 A2 (due to the circuitry involved via Q1 and Q2) due to the specific resonance effects present on each side (initiator and target) of the link in the setup.

7.2.2.1. *Optimal values of A1 and A2*

The optimal values of A1 and A2 are obtained by minimizing the values of the denominators of each of these terms, meaning that for: $x = y = 1$ – i.e. when the two initiator and target circuits have the same resonance frequency *and* are tuned to it – then we have:

$$|A1\ max| = Q1$$

$$|A2\ max| = Q2$$

Thus, and *only* in this case:

$$|n^2 \max| = k^2 Q1\,Q2 \quad n = k\sqrt{(Q1\,Q2)}$$

To continue: $n = k\,[\sqrt{[(L1\omega/R1) \times (L2\omega/R2)]}]$

so $n = k\,\omega\,\sqrt{(L1L2/R1R2)}$

Knowing that $M = k\,\sqrt{(L1L2)}$ this yields

$n = (M\,\omega)/\sqrt{R1R2}$

which gives us: $n^2 = (M\,\omega)^2/R1R2$

$M\,\omega = n\,\sqrt{(R1R2)}$

7.2.2.2. *Values of "n" and definitions of weak, critical and tight coupling*

By definition:

– when $0 < n < 1$ we say that the coupling is weak;

– when $n = 1$ we say that the coupling is critical;

– when $n > 1$ we say that the coupling is tight.

7.2.3. *In conclusion, an important point*

The energy transferred is, therefore, proportional to the *square of the coupling coefficient k* ($n^2 = k^2\,Q1Q2$), and it is as though *each resonance* – that of the generator (the initiator antenna) and that of the charge (the target) – amplifies the energy transfer *by a factor equal to the product of the quality factors Q1 and Q2 of their respective circuits*. Even more explicitly, the real value n^2 (or of course its root) involves:

– *first, the physical coupling coefficient k* (which depends only on the mechanical shapes of the coils, the distance between them and the nature of the magnetic environment (μ) separating the circuits, and is therefore independent of the internal technology of the initiators and targets);

– *second, the electrical quality factors Q1 and Q2 of the circuits* (which, for their part, are inextricably linked to the electrical technologies used in the initiators and targets).

Thus, it takes the form of a simple product between the physical coupling and a purely-technological amplification factor, which could also be called the merit factor. This mixed value is omnipresent in the study of coupled resonating circuits.

IN SUMMARY.– *The coupling index "n"* is, therefore, a parameter which *depends not only on the mechanics of the system by way of "k"*, but also on the *electrical setup* used by way of the quality coefficients Q1 and Q2 of the circuits present. Indeed, the values of the quality coefficients are linked to circuitry considerations (tuning, etc.) independent of the mechanics. With the same mechanics, k is only a function of Q.

By introducing the idea of the coupling index "n", we can use the resonances – at a definite frequency and when the two circuits are tuned to that same frequency – to obtain a considerable increase in the coupling factor between the initiator and the target – each of them resonant and separately playing the role of a coupling amplifier. When the physical and magnetic link is (sufficiently) weak (i.e. when the circuits are (sufficiently) far apart), the direct interpretation that is possible to give to this twofold amplification is that the resonance enables each circuit to couple better with the external field surrounding it.

Nevertheless, unlike "k", the value "n" is a complex number, and its amplitude is not necessarily less than one. Furthermore, the previous analysis is partial, because it does not take account of the value of the final charge (because for the instant, the charge circuit is open – of infinite impedance – and therefore equivalent to the condition of the threshold field of the integrated circuit for our target or tag). We will see that the behavior of such charged circuits is more complex because the concepts of tuning frequency and impedance matching are generally interwoven and dependent on the value of the physical coupling and the relative values of quality factors also when they are single-ended.

7.3. Identical coupled circuits, tuned to the same frequency

For didactic purposes and in the spirit of simplification, consider the diagram (Figure 7.6). It is made up of two strictly identical circuits R, L, C* (L1 = L2 = L; C1 = C2 = C; R1 = R2 = R) whose impedances Z are identical (*remember*: *optimum conditions of the coupling index, n*) and which are coupled via a mutual inductance M whose coupling coefficient k of the setup (k = M √ (L L) = M L) and of course less than 1. The value of k can be a rendered variable – for example, by altering the distance between the initiator and target.

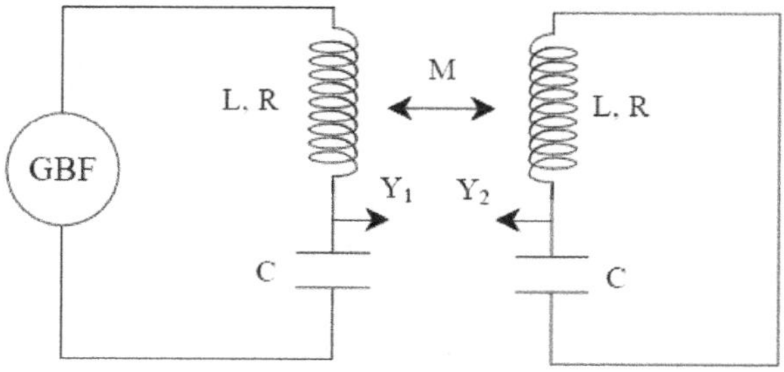

Figure 7.6. *Identical coupled circuits, tuned to the same frequency*

IMPORTANT NOTE.– This case is generally not actually the reality but, after wise calculations which we will see later on (different circuits, different tunings, etc.), we will see that it is always preferable to plan for this eventuality.

The equations of the two circuits (primary and secondary) are, therefore, as follows when we apply a voltage V to the setup:

$$V1 = Z\,I1 + j\,M\,\omega\,I2 = (R + j\,(L\omega - 1/C\omega)\,I1 + j\,M\,\omega\,I2$$

$$0 = Z\,I2 + j\,M\,\omega\,I1 = (R + j\,(L\omega - 1/C\omega)\,I2 + j\,M\,\omega\,I1$$

From this, it is easy to deduce the expressions of the currents I1 and I2:

$$I1 = \frac{Z \quad V1}{Z^2 + M^2\omega^2}$$

$$I2 = \frac{-j\,M\omega \quad V1}{Z^2 + M^2\omega^2}$$

Now, let us redraw the diagram so as to look specifically at the voltage V2 present at the terminals of the capacitor C at the secondary (in fact, when it is working as the threshold of the target's integrated circuit). In order to do so, we excite the primary by a sinusoidal voltage generator V1 (Figure 7.7).

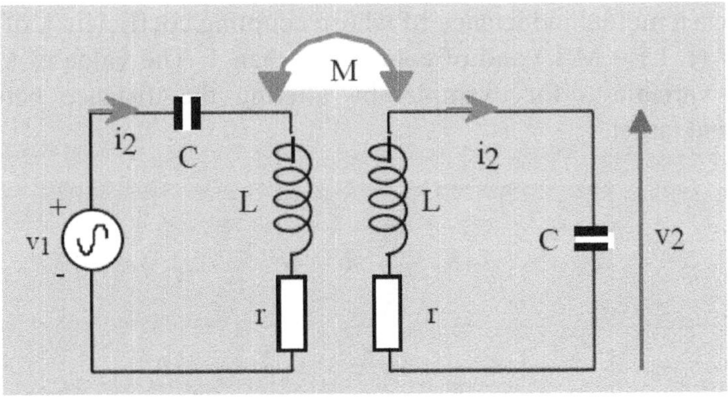

Figure 7.7. *Calculation of the voltage V2 at the terminals of the capacitor C*

7.3.1. *Transfer function, A(ω) = V2/V1, in terms of the voltage of the secondary*

Supposing that the applied voltage V1 arising from the integrated circuit of the initiator is constant, let us now calculate the expression of the voltage transfer function A = V2/V1 in order to find the value of the voltage V2 received by the target's integrated circuit.

When $Z = R + j(L\omega - 1/C\omega)$, if (similarly to the previous sections) we express this diagram in equation form, it gives us the following group of relations:

$$V1 = Z\,I1 + j\,M\,\omega\,I2$$

$$-j\,M\,\omega\,I1 = Z\,I2$$

$$V2 = I2/j\,C\,\omega$$

V2 *…at the terminals of the secondary's capacitor and therefore of the integrated circuit*

With these equations, we can easily deduce the transfer function A(ω) = V2/V1 of the setup:

$$A(\omega) = \frac{V2}{V1} = \frac{M}{C} \frac{1}{(Z^2 + M^2\,\omega^2)}$$

NOTE.– The function A(ω) represents (nearly) the relation between the portion of voltage induced by the external field H (in A/m) coming from the initiator (the reader) and the applied voltage V2 (in volts) at the terminals of the target integrated circuit just at its operating threshold. Almost finally, there have been numerous simplifications, which we will now correct.

Let us return to the expression of A(ω) and introduce the relations with the coupling index $n = M\,\omega/R$, where $Q = 1/RC\omega$:

$$A(\omega) = \frac{V2}{V1} = \frac{n}{R\,C\,\omega}\;\frac{1}{(1 + j\,X/R)^2 + n^2}$$

$$A(\omega) = \frac{V2}{V1} = \frac{n}{(1 + j\,X/R)^2 + n^2}\,Q \text{ regardless of the value of } \omega$$

7.3.2. *Transmission coefficient "Kt"*

By definition, we set Kt = transmission coefficient:

$$Kt = \frac{1}{(1 + j\,X/R)^2 + n^2} \text{ … which is a value of a complex variable.}$$

$$A(\omega) = \frac{V2}{V1} = -\;Kt\;Q \quad \text{regardless of the value of } \omega.$$

At the resonance, when $X = 0$, the value of the transmission coefficient "Kt" is simplified and becomes:

$$Kt = \frac{n}{1 + n^2}$$

The (simple) study of the function Kt = f(n) shows that it passes through a maximum (and thus the current I2 does too), equal to 0.5 at the so-called critical coupling level for $n = 1$.

Thus: $$I2\ max = -j\,\frac{V1}{2\,\sqrt{R1R2}}$$

The graph of Kt = f(n) is presented in Figure 7.8.

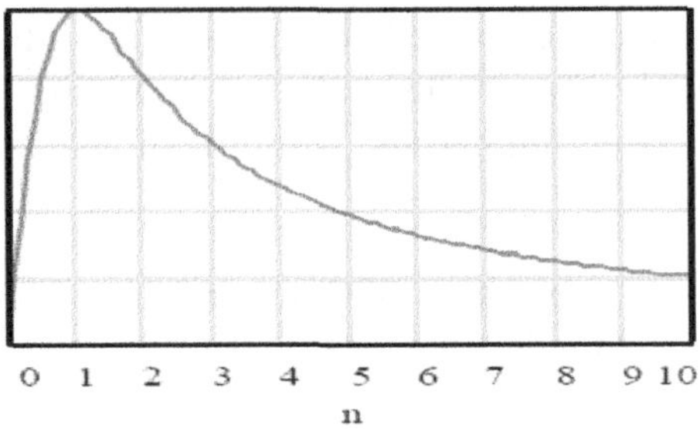

Figure 7.8. *Graph of Kt = f(n)*

where we define:

	n	Coupling	Kt
When	$0 < n < 1$	Weak	Increases from 0 to 0.5
	$n = 1$	Critical	Is at its max = 0.5
	$n > 1$	Tight	Decreases from 0.5 to 0

Table 7.2. *Values and appelations of the coupling index*

7.3.3. *In summary*

All these peculiarities define, in total, five different important appellations.

		=	=	=
Mutual perfect (in Henry)	"M"	$\sqrt{L1L2}$		
Coupling coefficient	"k"	$M / \sqrt{L1L2}$		
Coupling index	"n"	$M\,\omega / \sqrt{R1R2}$	$k\,\omega\,\sqrt{(L1L2 / R1R2)}$	$k\,\sqrt{Q1Q2}$
Transmission coefficient	"Kt"	$n / (1+n^2)$		
Transfer function	"A(ω)"	$-\,Kt\,Q$		

Table 7.3. *Summary of the main parameters*

7.3.3.1. *Continued example, with several coupling measurements*

As in the previous example, the measurement is performed with two NFC devices: one the initiator, and the other the target:

– for example, at the terminals of the initiator antenna, we inject a voltage V1 of 2.5 V;

– we measure the no-load voltage V2 at the terminals of the coil in the target antenna.

	Equations	Units	Initiator	Target
R		Ω	1	7
L		μH	0.7	0.53
X	= L ω	Ω	60.75	45.6
Q	= X/R		Q1= 57.3	Q2= 6.5
Measured voltage		V	V1= 2.5	V2 = 0.38
K	= (V2 / V1) [√(L1/L2)]			0.1687
N	= k * √(Q1 *Q2)			3.25
Kt	= n / (1 + n²)			0.28
A(ω)	= Kt √Q1*Q2			5.58
M	= k √L1 L2	nH		103

Table 7.4. *Examples of coupling measurements*

We obtain all the coefficients in a series of a few measurements. Mission accomplished.

7.3.3.2. *Calculation of the voltage ratio V2/V1 f(ω) = A(ω)*

Let us now look at the evolutions of the voltage V2 as a function of the frequency.

This calculation enables us to find the voltage present at the terminals of the capacitor C of the secondary … so the voltage which will be developed at the terminals of the integrated circuit of the target just at the threshold (capacitor not heavily loaded by the input resistance of the integrated circuit).

For this, irrespective of the value of ω, without approximation, we set:

$$\delta = (\omega - \omega 0)/\omega 0 = (f - f0)/f0 = \Delta f/f0$$

or indeed:

$$\omega = \omega 0\ (1+\delta)$$

The parameter "δ" represents what is known as "dissonance". The expression of Z then becomes:

$$\begin{aligned} Z &= R + j\,(L\omega - 1/C\omega) \\ &= R + j\,L\omega 0\ [(1+\delta) - 1/(1+\delta)] \end{aligned}$$

or indeed:

$$Z = R\,(1 + j\,\frac{L\,\omega 0}{R} \times \frac{\delta\,(2+\delta)}{1+\delta})$$

$$Z = R\,(1 + j\,Q\ \delta\,\frac{(2+\delta)}{1+\delta})$$

, the exact value of Z regardless of the value of ω.

Let us set a new variable h':

$$h' = Q\,\delta\,(\frac{(2+\delta)}{1+\delta})$$

We find Z = R (1 + j h') and by substituting that value of Z back into A(ω), we find:

$$A(\omega) = \frac{V2}{V1} = \frac{M}{C}\ \frac{1}{(R^2(1+j\,h')^2 + M^2\omega^2)}$$

$$A(\omega) = \frac{V2}{V1} = \frac{M}{C}\ \frac{1}{R^2(1+j\,h')^2 + n^2\,R^2)}$$

Let us, once again, use relation n = M ω/R:

$$A(\omega) = \frac{V2}{V1} = \frac{n}{R\,C\,\omega}\ \frac{1}{(1+j\,h')^2 + n^2}$$

In the knowledge that $\omega = \omega0\ (1 + \delta)$:

$$A(\omega) = \frac{V2}{V1} = \frac{n}{R\,C\,\omega0\,(1+\delta)} \times \frac{1}{(1 + j\,h')^2 + n^2}$$

and also, $Q = 1/RC\omega o$, we finally obtain:

$$A(\omega) = -\frac{n\,Q}{(1+\delta)\,[(1 + j\,h')^2 + n^2]} = \frac{n\,Q}{(1+\delta)\,[\,(1 + n^2 - h'^2) + j\,2\,h']} = V2/V1$$

Hence, the module of the transfer function $A(\omega) = V2/V1 = f(\omega)$:

$$|A| = \frac{n\,Q}{(1+\delta)\,\sqrt{[\,(1 + n^2 - h'^2)^2 + (2\,h')^2\,]}}$$

by substituting:

$$h' = Q\,\delta\,(\frac{(2+\delta)}{1+\delta})$$

we obtain the equation without approximation:

$$|A| = \frac{Q\,n\,(1+\delta)}{\sqrt{[(1+\delta)^2 + n^2(1+\delta)^2 - Q^2\delta^2\,(2+\delta)^2)^2 + 4\,Q^2\,\delta^2(\,2+\delta)^2\,]}}$$

7.3.4. *Operation in the vicinity of the resonance frequency f0*

In the case of $\delta < 1$ and $\delta^2 << 1$ and the initial value of h' is simplified and is reduced to a new variable known as "h", which has the value:

$h = 2\,Q\,\delta$ which thus represents the new frequency variable.

$$Z = R\,(1 + j\,Q\,2\,\delta)$$

Similarly, beginning with the general equations (with no approximation) for |A|, recapped below:

$$A(\omega) = \frac{n\,Q}{(1+\delta)\,[(1 + j\,h')^2 + n^2]} = \frac{n\,Q}{(1+\delta)\,[\,(1 + n^2 - h'^2) + j\,2\,h']} = V2/V1$$

$$|A| = \frac{n\,Q}{(1+\delta)\,\sqrt{[\,(1+n^2-h'^2)^2+(2\,h')^2\,]}}$$

we can easily obtain the relations of the transfer function A(ω) = V2/V1

$$A = \frac{-\,n\,Q}{(1+j\,h)^2+n^2} = \frac{n\,Q}{(1+n^2-h^2)+j\,2\,h}$$

and its modulus, the equation with approximation where $\delta << 1$

$$|A| = \frac{n\,Q}{\sqrt{[\,(1+n^2-\ h^2)^2+(2\,h)^2\,]}}$$

or indeed, the equation with approximation where $\delta << 1$

$$|A| = \frac{n\,Q}{\sqrt{[\,(1+n^2-\ 4\,Q^2\delta^2)^2+16\,Q^2\delta^2\,]}}$$

7.3.4.1. *Frequency response of A(ω)*

In the vicinity of the resonance frequency f0, with h close to zero, the modulus of the transfer function |A| is given by the equation below:

$$|A| = \frac{n\,Q}{\sqrt{[(1+n^2-h^2)^2+(2\,h)^2]}} = f\,(n,\,h) \text{ modulus of the transfer function } A(\omega)$$

and, only at the resonance (on the carrier), so for $\omega^2 = 1/(LC)$, X = 0, h = 0

$$A(\omega) = \frac{V2}{V1} = \frac{n\,Q}{1+n^2} = Kt\,Q$$

The function A(ω) thus has a shape similar to Kt as a function of "n" and touches a maximum equal to Q/2 for n = 1.

7.3.4.2. *Extreme values (min. and max.) of the transfer function*

To determine the abscissa values of the extremes of the transfer function |A|, we derive the one above (which is of type "a/√u") in relation to the frequency variable h (= 2 Q (f – fo)/fo):

$$|A| = -\frac{n\,Q}{\sqrt{[(1+n^2-h^2)^2+(2\,h)^2]}} = f(h)$$

After expansion of the denominator after simplifications:

$$|A| = -\frac{n\,Q}{\sqrt{[(h^2(h^2-2(n^2-1)) + (n^2+1)^2]}} = \frac{u}{v} = f(h)$$

The derivative dA/dh = A'(h) of A in relation to h is equal to:

$$A'(h) = \frac{nQ}{2} \times \frac{[[(h^2(h^2-2(n^2-1)) + (n^2+1)^2]\text{^}3/2] \times [2h\,(h^2-(n^2-1)]]}{v^2}$$

This takes a value of 0 when the factor $[2h\,(h^2 - (n^2-1)] = 0$, and we then find three values of h that give a zero value to the derivative:

$h = 0$

i.e. for f = f0:

– corresponds to a unique maximum of |A| for values of $0 < n < 1$ or equal to 1;

– corresponds to a minimum of |A| for values of $n > 1$.

$h = +\sqrt{(n^2-1)}$ when $n > 1$ these two values correspond to two lateral differential maxima in relation to the central minimum of 1.

$h = -\sqrt{(n^2-1)}$

Obviously, these last two maxima can only exist *on condition that n > 1*, known as *"tight coupling"*.

Hence, there is a particular value of the coupling index *corresponding to n =1*, known as *"transitional coupling"* or *"critical coupling"*, at which the lateral maxima disappear, forming only one.

Conversely, for n <1, known as *weak coupling*, we now only have the solution h = 0, which becomes a unique central maximum.

Figures 7.9–7.11 illustrate an example of results of variations in modulus of the transfer function |A| = V2/V1 in f(ω) obtained as a function of the variations in "h" (thus with ω) varying from –5 to 5, and simultaneously with variations in the coupling index "n", which varies in two stages: from 0.5 to 1 (Figure 7.9), and then from 1 to 3 (Figure 7.10) for tuned circuits which each have, for example, a quality coefficient Q equal to 100.

7.3.4.2.1. Weak coupling: n < 1

Figure 7.9 indicates that for a value of the coupling index n < 1 (0.5–1), the response curve has only one maximum point, which corresponds to f = f0, but whose value is less than Q/2.

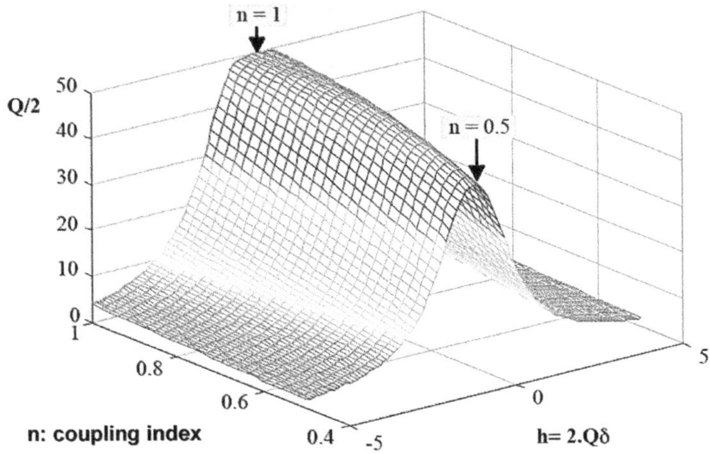

Figure 7.9. *n < 1 – weak coupling. For a color version of the figure, see www.iste.co.uk/paret/antenna.zip*

7.3.4.2.2. Critical coupling: n = 1

It can be seen from the same figure that, for n = 1, the critical/transitional coupling yields a curve with a flattened maximum, Bp = √2 fo/Q, which then plummets rapidly, giving a better bandpass filter.

7.3.4.2.3. Tight coupling: n > 1

Figure 7.10 shows that, for values of the coupling index n > 1 (from 1 to 3), the response resonances differ from the resonance eigenfrequency f0 (where h = 0) of the serial RLC circuits, and are more acute, whose value of summits which remains constant is equal to Q/2 (see a little later on).

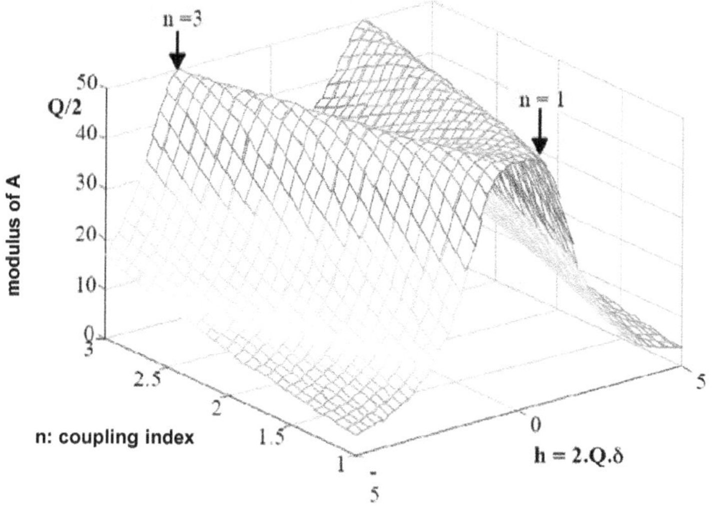

Figure 7.10. *Tight coupling: n > 1. For a color version of the figure, see www.iste.co.uk/paret/antenna.zip*

7.3.4.2.4. In summary

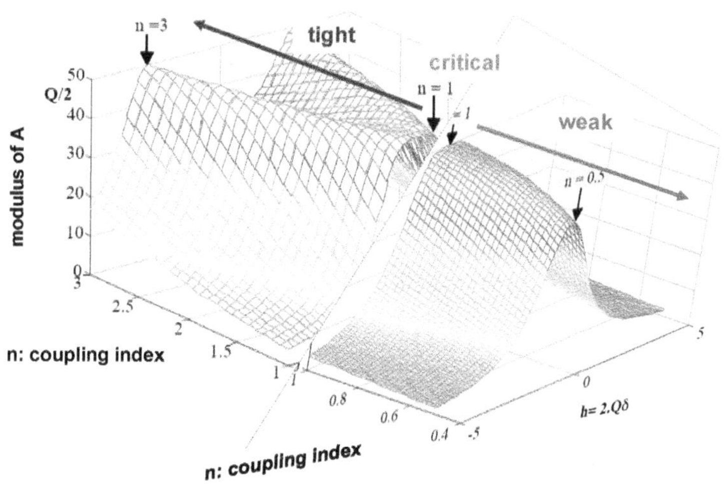

Figure 7.11. *Overview of coupling indices. For a color version of the figure, see www.iste.co.uk/paret/antenna.zip*

In *NFC ISO 18092* (and *ISO 14443*), to successfully pass the modulation frequencies of the sidebands of the two subcarriers situated at +/– 848 kHz of the carrier *fo = 13.56 MHz*, we need to have a total bandwidth of the order of Bp = 1.356 MHz around fo, which are the two peaks of the curve A(ω), essentially at 12.882 and 14.238 MHz. Thus, we must have:

	Bp	= 1.356 MHz
	f - fo = Δf	= Bp/2
		= 1.356/2 MHz
		= 0.678 MHz
Consider:		δ = (f – fo)/fo
	δ	= 0.05
With:	Q	Q = √ Q2Q3 imposed by the bits rates and the pause in ISO 14443
		= ~ 20
	h	= 2 Q δ
		= 2 × 20 × 0.05
	h	= 2

In addition, the abscissa values of the peaks are given by the relation:

	$h = +\sqrt{(n^2 - 1)}$
where:	$h^2 = (n^2 - 1)$
Consider:	$4 = n^2 - 1$
Hence:	$n^2 = 3$
Let:	$n = 1.732$
This gives us:	
Kt	$= n/(n^2+1)$
	$= 1.732/4$
Kt	$= 0.433$
A(ω)	= Kt Q
	= 0.433 × 20
A(ω)	= 8.66

Box 7.1. *Example of NFC systems – concrete and realistic*

7.3.4.3. *Values of the central extremum of |A| regardless of the value of "n"*

If, regardless of the value of "n", we look at the value of the *central extremum* of the function |A| – i.e. for fo, *when h = 0*, its value:

$$|A| = \frac{n\,Q}{\sqrt{[(1 + n^2 - h^2)^2 + 4\,h^2]}}$$

is then as follows, for h = 0:

$$|A\_fo| = \frac{n\,Q}{1 + n^2} = Kt\,Q = f(n)\ldots \text{ regardless of "n", so dependent upon "n"}$$

The rather special shape of the values of that *central extremum* is shown below:

– from n = 0 to n = 1, this extremum corresponds to a maximum whose value is increasing;

– for n = 1, the extremum touches a maximum maximorum equal to Q/2;

– and then from n = 1 to n = infinite, that extremum corresponds to a trough in the curve, where its value decreases with n. *This will be the origin of the loading effect due to the coupling effect.*

By way of example, the curve in Figure 7.12 shows the variations of |Amax| as a function of the coupling n for circuits with an initial Q of 100.

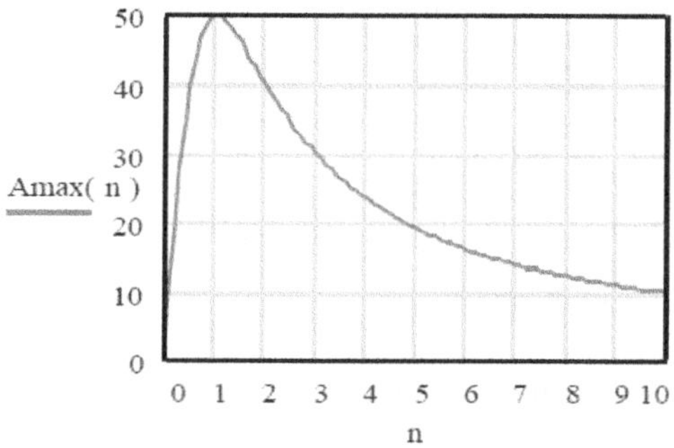

Figure 7.12. *Variations of |Amax| = f(n) for circuits with initial Q = 100*

NOTE.–

– Ultimately, for values of n = 0.7 to n = 2, we are in the range of 80% of the possible maximum, which is equal to Q/2;

– in the above example, where n = 1.732, A = 0.433 Q, and we are at 0.433/0.5 = 86.6% of the maximum of A (the fact of not being exactly at the optimum point of coupling n = 1 has cost around ≈ 14%).

7.3.4.4. *Values of the central extremum of |A| when n → 0 (very weak coupling)*

When n < or = *at 1*, the value of the modulus of |A| has only one (single) maximum |Amax|. If, now, the value of "n tends toward 0" and thus n is small in relation to h, the equation:

$$|A| = - \frac{n\,Q}{\sqrt{[(1 + n^2 - h^2)^2 + 4\,h^2]}}$$

becomes

$$|A\_o| = - \frac{n\,Q}{\sqrt{[(h^2 + 1)^2]}} \quad \text{for } n \to 0$$

so

$$|A\_o| = - \frac{n\,Q}{(h^2 + 1)} \quad \text{for } n \to 0$$

7.3.4.4.1. Bandwidth at –3dB of the setup as n → 0

For n → 0, let us find the values of h giving a bandwidth at –3dB:

$$\frac{n\,Q}{h^2 + 1} = \frac{1}{\sqrt{2}} \times \frac{n\,Q}{1}$$

Thus

$$\begin{aligned} h^2 &= \sqrt{2} - 1 = 0.414 \\ h &= \sqrt{(\sqrt{2} - 1)} = 0.643 \\ &= 2\,Q\,\delta \\ &= 2\,Q\,(f - fo)/fo \\ 2(f - fo) &= 0.643\ fo/Q \end{aligned}$$

Hence, for n → 0

$$\text{Bp at } -3\text{dB} = 0.643 \frac{fo}{Q}$$

7.3.4.5. *Values of the central extremum of |A| when n = 1*

The value of "n" = 1 is a very specific point of the function |A|. This is "critical coupling" or "transitional coupling", and the place where three extrema come together into one.

In this case, "n =1":

– the value of the common central extremum to the modulus of |A| passes through a maximum |A max| whose value is equal to Q/2, which is half of what we obtain with a unique resonating R L C serial circuit;

– for n = 1, the value of |A| as a function of "h" becomes:

$$|A| = \frac{n\,Q}{\sqrt{[(1 + n^2 - h^2)^2 + 4\,h^2]}}$$

$$|A| = -\frac{Q}{\sqrt{[(4 + h^4]}}$$

For n = 1, let us now search for the values of "h" for which the *value of the modulus of* $|A|$ *is reduced by* $1/\sqrt{2}$ (i.e. *–3dB)* of its maximum amplitude. Therefore, we need to solve the equation:

$$|A| = \frac{Q}{\sqrt{[(4 + h^4)]}} = \frac{1}{\sqrt{2}} \times \frac{Q}{2}$$

$$\sqrt{[(4 + h^4)]} = 2\sqrt{2}$$

Let us square this:

$$4 + h^4 = 8$$

so:

$$h = +- \sqrt{2}$$

In addition, remembering that:

$$h = 2 Q \delta$$

$$\delta = \frac{+\sqrt{2}}{2Q} = \frac{f - fo}{fo}$$

we find $f1' = fo\left(1 - \frac{\sqrt{2}}{2Q}\right)$ and $f2' = fo\left(1 + \frac{\sqrt{2}}{2Q}\right)$

Fairly flat because it is of form $Q/\sqrt{(4 + h^4)}$. The bandwidth (f2 – f1) at –3dB is therefore:

$$n = 1\ (f2 - f1) = Bp \text{ at -3dB} = \sqrt{2}\frac{Fo}{Q} = 1.414\frac{fo}{Q}$$

7.3.4.6. *Values of the two lateral maxima of |A| when n > 1*

In the case that the coupling index n > 1, it is easy to show (see below) that *irrespective of the values of n (n > 1)* the ordinate values of the two lateral maxima are and remain constant and correspond to the two successive resonances on frequencies f1 and f2 that are different from the eigenfrequencies f0 which are always on either side, as we showed earlier.

Let us calculate the values of |A| for h = + and - $\sqrt{(n^2 - 1)}$... (which is completely indifferent because h appears in the square). Thus, we obtain:

$$|A| = \frac{n\,Q}{\sqrt{[(1 + n^2 - h^2)^2 + 4\,h^2]}}$$

$$|A| = \frac{n\,Q}{\sqrt{((2)^2 + 4n^2 - 4)}}$$

$$A1\,(f1) = A2\,(f2) = |Amax| = \frac{Q}{2} \text{ regardless of the values of “n” >1.}$$

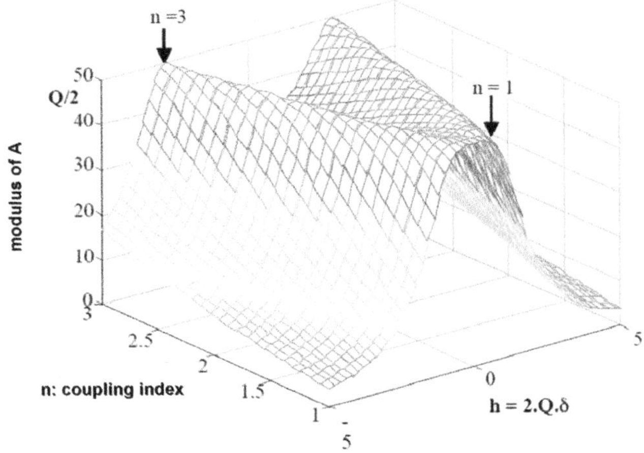

Figure 7.13. *Values of the minimum and the two lateral maxima of |A| when n > 1. For a color version of the figure, see www.iste.co.uk/paret/antenna.zip*

Now that we have determined the general equation for the variations of |A| = f(n, h):

$$|A| = \frac{n\,Q}{\sqrt{[(1 + n^2 - h^2)^2 + 4\,h^2]}}$$

and, also, for a given coupling index "n" greater than 1, the literal values:

– first, two maximum amplitudes:

$$A1\ (f1) = A2\ (f2) = |Amax| = \frac{Q}{2} \text{ regardless of the values of "n" >1.}$$

– second, the minimum in the trough:

$$|A \min| = \frac{n\,Q}{1 + n^2} = f(n)\ldots \text{ and thus dependent upon "n"}$$

Let us now look for the particular remarkable values of the curve, for two examples.

In the case where n > 1, the peaks of the lateral maxima, differential in relation to the central minimum, via the relations $h = 2\ Q\ \delta$ and $\delta = (f - fo)/fo$ correspond to the following two frequencies which are roots of the equation:

$$h \quad = +- \sqrt{(n^2 - 1)} \quad = 2\ Q\ ((f - fo)/fo)$$

$$f1 = fo\ (1 - \frac{\sqrt{(n^2 - 1)}}{2Q}) \text{ and } f2 = fo\ (1 + \frac{\sqrt{(n^2 - 1)}}{2Q})$$

7.3.4.6.1. Example 1

For a given/known value of "n" greater than 1, let us find the values of h1' and h2' (or frequencies f1' and f2') for which the attenuation values on the sides are identical to that which is present in the central trough. This entails solving the "h" equation with a given value of "n":

$$|A| = - \frac{n\,Q}{\sqrt{[(1 + n^2 - h^2)^2 + 4\,h^2]}} = \frac{n\,Q}{n^2 + 1}$$

In order for that equation to be verified, it is necessary for the denominator $\sqrt{(a^2 + b^2)} = n^2 + 1$, so indeed:

$$h^2\ (h^2 - 2(n^2 - 1)) = 0$$

This equation can have three roots, h0' = 0 and h1' and h2' = + and - $\sqrt{(2(n^2 - 1))}$.

As the peaks of the curve take place at $h = \sqrt{(n^2 - 1)}$ … the new values h1' and h2' are at "h $\sqrt{2}$", as h = 2 Q δ, so δ' = δ $\sqrt{2}$ and the associated frequencies are therefore (Figure 7.14):

$$h' = \pm \sqrt{(2(n^2 - 1))} = 2\,Q\,((f - fo)/fo)$$

$$f1' = fo\left(1 - \frac{\sqrt{(2(n^2 - 1))}}{2Q}\right) \text{ and } f2' = fo\left(1 + \frac{\sqrt{(2(n^2 - 1))}}{2Q}\right)$$

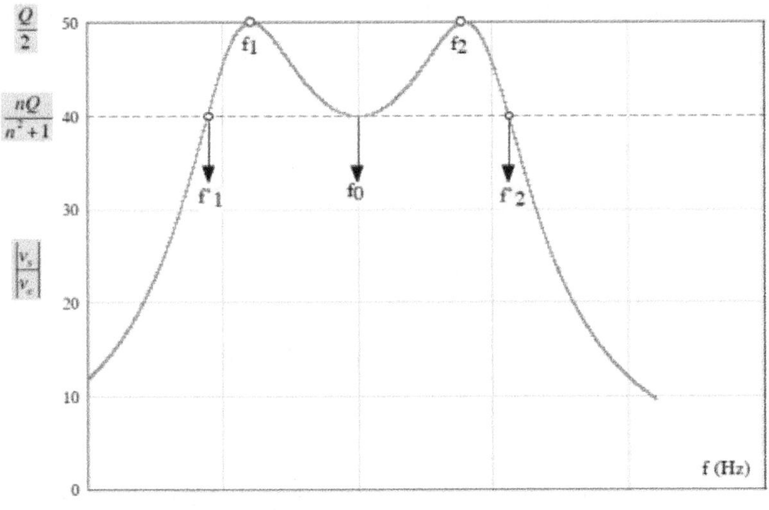

Figure 7.14. *Minimum and maximum values on the curve*

The distance (f2' – f1') is:

$$n > 1 \quad f2' - f1' = \sqrt{(2(n^2 - 1))}\,\frac{fo}{Q} = (\sqrt{(n^2 - 1)})\,.\,\sqrt{2}\,\frac{fo}{Q}$$

The amplitudes in f1' and f2' are equal to $nQ/(n^2 + 1)$. The relative attenuation of the trough in relation to the peaks is equal to $(nQ/(n^2 + 1)/(Q/2) = 2n/(n^2 + 1)$, which is in dB $\log 10\,(2n/(n^2 + 1))$.

7.3.4.6.2. Example 2: bandwidth at –3 dB

Let us calculate the value of the coupling index "n" *for which the above voltage ratio is equal to –3 dB (a ratio of 1/√2)* and gives the value of the bandwidth … but with a trough:

$$\frac{2n}{(n^2 + 1)} = \frac{1}{\sqrt{2}}$$

i.e. the following second-degree equation: $n^2 - n\,2\sqrt{2} + 1 = 0$; the roots of this equation are: n = 2.414 and 0.414.

As we are in the case where n > 1, we only conserve n = 2.414.

The peaks are, of course, at a maximum amplitude of Q/2, but the center is (Q/2)/√2 (either 3 dB below, or "the peaks/1.414") … which means that on the carrier, we can lose 41.4% of the quality of the remote supply of the target in relation to n = 1. What we gain in band, we lose in carrier (Figure 7.15).

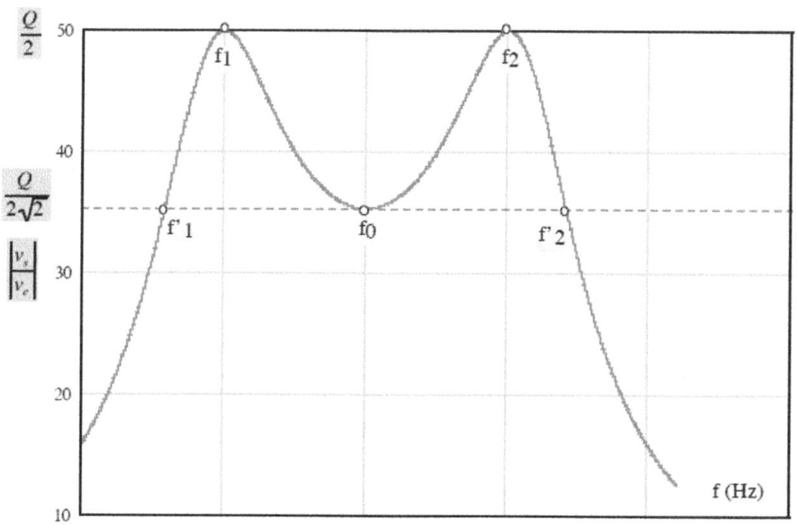

Figure 7.15. *What we gain in band, we lose in carrier*

The bandwidth (f2' – f1') at –3 dB is:

$$Bp \text{ at } -3\text{dB} = f2' - f1' = \sqrt{(2\,(n^2 - 1))}\ \frac{fo}{Q}$$

$$\text{where } n = 2.414\ Bp \text{ at } -3\text{dB} = 3.10\ \frac{fo}{Q}$$

IN SUMMARY.– The classic, conventional two-dimensional (2D) figure of the example shown in Figure 7.16 summarizes the curves for all the responses presented here, with no detail or didactic aspects.

EXAMPLE.– At 10.7 MHz – f0 = 10.7 MHz, C = 220 pF, L = 1 µH, r = 0.68 Ω and Q = 100 and a coupling index n ranging from 0.2 to 2.4.

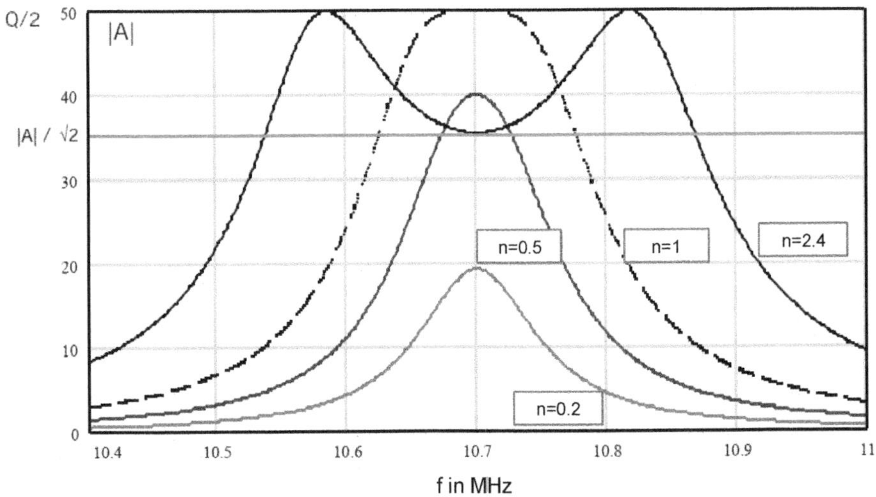

Figure 7.16. *Example of 2D representation. For a color version of the figure, see www.iste.co.uk/paret/antenna.zip*

7.3.4.7. *Table summarizing couplings*

<table>
<tr><td rowspan="2"></td><td>Coupling index</td><td>Type of coupling</td><td>Selectivity</td><td>Bandwidth
Bp = 2 Δf
@ - 3 dB</td><td colspan="2">Example: coupling coefficient at same Bp
f0 = 13.56 MHz
bit rate = 106 kbits/s</td><td>Energy</td></tr>
<tr><td>n</td><td></td><td>S</td><td>Bp</td><td>Bp @ 3 dB = 386 kHz (iso)
Q</td><td>S</td><td></td></tr>
<tr><td>Single-antenna</td><td></td><td>No coupling</td><td>$\sqrt{(1 + 4 Q^2(\Delta f / f0)^2)}$
$\sqrt{(1 + 4 Q^2(Bp^2 / 4 f0^2))}$</td><td>f0 / Q</td><td>35</td><td>1.411</td><td></td></tr>
<tr><td rowspan="6">Two coupled antennas</td><td>0</td><td>No coupling</td><td colspan="5"></td></tr>
<tr><td>near to 0</td><td>Very weak</td><td></td><td>0.643 f0/Q</td><td></td><td></td><td></td></tr>
<tr><td>0 to 1</td><td>Weak</td><td>$1 + 4 Q^2 (\Delta f / f0)^2$
$1 + 4 Q^2 (Bp^2 / 4 f0^2)$</td><td></td><td>22.5</td><td>1.410</td><td>at Q = 35
Boosted</td></tr>
<tr><td>near to 1</td><td>Critical</td><td>$\sqrt{(1 + 4 Q^4 (\Delta f / f0)^4)}$
$\sqrt{(1 + Q^4 (Bp^4/4/f0^4))}$</td><td>1.414 f0/Q</td><td>49.67</td><td>1.4139</td><td></td></tr>
<tr><td>>1</td><td rowspan="2">Tight</td><td></td><td></td><td></td><td></td><td></td></tr>
<tr><td>2.414</td><td></td><td>3.10 f0/Q</td><td></td><td></td><td></td></tr>
</table>

Table 7.5. *Summary of couplings*

EXAMPLES OF MEASUREMENTS.–

Measurements		Initiator + target	
		Initiator	Target
Number of turns		2	4
R ()	Ω		
L (	µH)	2	4
Lo		1	1
Q = X/R		25	25
R	cm	3.5	
s2	mm^2		$70 \times 43 = 3000$
Troughs/peaks (Q/2 / √2) / Q/2	%	70	
$n = k * \sqrt{(Q1 * Q2)}$		2.5	
$k = n / \sqrt{(Q1 * Q2)}$		0.1	
$Kt = n / (1 + n^2)$		0.34	
$A(\omega) = Kt \sqrt{Q1 * Q2}$		8.62	
$M = k \sqrt{L1\ L2}$	nH	$0.1 * \sqrt{(2*4)} = 282$	
$k = (V2 / V1) [\sqrt{(L1/L2)}]$		$0.38/2.6 [\sqrt{(0.7127/0.535)}] = 0.1687$	$1.58/2.6 [\sqrt{(0.7127/1.32)}] = 0.447$
Distance		$k = (\mu \times \frac{r^2}{2(r^2 + d^2)^{3/2}}) \times s2] \times \sqrt{(\frac{1}{L1o \times L2o})}$ Hence "d"	

Table 7.6. *Examples of measurements*

8

The Initiator–Target Couple and the Loading Effect

Very often, a near-field communication (NFC) system operates with tight or very tight coupling, because the desired applicative distance of communication is small for secure "*Touch and Go*" or "*Touch and Confirm*" applications. Such is the case, for example, when an "NFC target", a mobile phone in card emulation mode, is read by an mobile Point of Sale (mPOS) initiator – see photos in Figure 8.1. The overvoltage present at the target due to A(ω) is then subtracted from that of the initiator, as a function of the value of its quality factor Q and its resonance eigenfrequency.

Figure 8.1. *Reading of an "NFC device target mobile phone in card emulation mode" on an initiator such as a POS or an mPOS*

8.1. Loading effect by coupling

All of this explains why (very) often, in passive mode (retro-modulation) – e.g. application in chip-card emulation mode by a mobile phone or tablet on a POS or mPOS – while the current trough in the initiator deepens and becomes more marked, the bumps in the frequency response separate in accordance with the relations set out above, the radiated field is reduced, and the applicative distance is greatly reduced. Indeed, not only is the no-load magnetic field produced by the initiator not very significant (because the POS/mPOS devices run on battery). In order for it to work, we must reduce the distance, bringing the target closer to the initiator in order to have a more intense field H because of an effect known as the *loading effect arising from coupling* because:

– the value of the no-load field present is again reduced;

– therefore, we need to bring the target closer to the initiator again;

– therefore, the coupling is increased again;

– thus, the response at the center becomes even deeper;

– hence, the current emitted by the initiator is reduced;

– hence, the field emitted by the initiator is also reduced;

– so, it is no longer possible to (remote) power the target properly, etc.

The more the signals are deformed, the less well the data will pass.

Here, then, is the full situation.

NOTE.– This effect is far less pronounced in the case of an ATM application, because the initiator antenna is situated a few centimeters away from the face (see Paret book 2) and thus the loading effect because of coupling due to the NFC target mobile phone is slight or less pronounced ... but, we need a stronger field because the distance is greater.

Following the detailed examination of the voltage V2 (voltage at the secondary at the target antenna), we will go on to look at the current I1 in the primary – i.e. the current in the initiator antenna.

8.2. Coupled tuned antennas in terms of the primary current

To present an image of the discussion topic of the previous section, let us again express the equations for the setup represented by "*two tuned primary and secondary circuits with inductive coupling*" (an initiator and a target) – see Figure 8.2 – but not necessarily on the same frequencies.

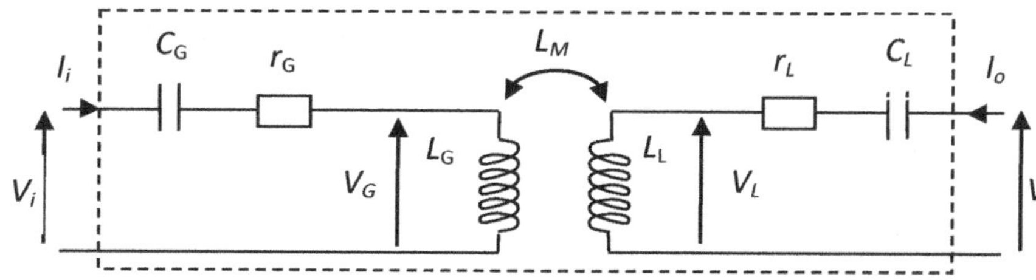

Figure 8.2. *Tuned primary and secondary circuits with inductive coupling*

If "E" (*Vi* in the figure) is the voltage present at the terminals of the antenna circuit of the generator (initiator), taking account of the potential differences induced on either side, because of the circulation of the primary and secondary currents, then using the mesh laws, we can write the following equations for both parts of the setup:

– E = voltage generator;

– M = mutual induction between the coils L1 and L2;

– R1 = serial resistance of the primary (R1 of the active antenna of the reader)

= (r1 of the base station antenna + rho internal resistance of the generator);

– R2 = total equivalent serial resistance of the secondary (the passive antenna)

= (antenna resistance + load + equivalent environment);

– X1 = L1ω – (1/C1ω) reactance of the primary;

– X2 = L2ω – (1/C2ω) reactance of the secondary;

– M mutual inductance (expressed in Henry) between the primary and secondary.

8.2.1. *Primary (initiator) non-loaded (no target within the field)*

To begin with, we quantify the value of the current I1o circulating in the circuit of the *initiator*, with "no load", in the absence of any targets within its field ("non-loaded" case). To do so, we write the simple equation relative to the mesh of the primary (initiator) in the initiator/target setup, with the target being absent.

$$[R1 + j(X1)]\ I1o = E$$

In this equation:

– R1 represents the equivalent serial value of all the resistances (generator, tuned circuit, matching circuit, etc.);

– X1 represents the equivalent total reactance of the components present in the primary of the setup (including, if need be, the impedance matching circuits such as "capacitive impedance bridges", etc.). Sooner or later, this reactance is written in the form $(L1\omega - 1/C1\omega)$ and $(\omega 1o)^2 = 1/L1C1$ – a value which, in principle, is not necessarily equal to the setups working frequency ωf.

NOTE.– In most RFID/contactless/NFC application, with no load:

– the initiator is normally tuned to the working frequency (13.56 MHz), and in this case $\omega 1o = \omega f$ and at that frequency $X1 = 0$ and the impedance represented by the initiator is real, with the value of R1;

– furthermore, to provide the maximum power to the initiator antenna, the impedance of the generator is normally matched via a "bridge" capacitive matching circuit, but we can turn it into an equivalent circuit;

– we can also express the whole of the initiator circuit using the following equivalent diagram:

generator = R

load = R

$L\,C\,\omega 1o^2 = 1$ at the resonance $V = 0$ – short-circuit

8.2.2. *Primary (initiator) with a load (presence of target(s) in the field)*

Let us write the equations for the two primary and secondary meshes:

$$E - j\,M\omega\,I2 = I1\,(R1 + j\,X1) \quad [8.1]$$

$$-j\,M\omega\,I1 = I2\,(R2 + j\,X2) \quad [8.2]$$

8.2.2.1. *View of the primary*

On the primary side, let us examine what an initiator circuit (active antenna) will perceive in the presence of a tuned target (passive antenna). To do this, we eliminate I2 from the second equation:

$$I2 = -j\,M\omega\,I1/(R2 + j\,X2)$$

We can substitute its value back into the first equation:

$$E - j\,M\omega\,I2 = I1\,(R1 + j\,X1)$$
$$E = I1\,[(R1 + j\,X1) + (M^2\omega^2/(R2 + j\,X2)]$$

However:

$$\frac{M^2\omega^2}{R2 + j\,X2} = \frac{M^2\omega^2\ (R2 - j\,X2)}{(R2 + j\,X2)\,(R2 - j\,X2)} = [R2\,\frac{M^2\omega^2}{R2^2 + X2^2} = j\,(X2\,\frac{M^2\omega^2}{R2^2 + X2^2})]$$

so, we have the final equation giving the relation between the potential difference E applied to the initiator antenna and the current in the primary I1 circulating in it, all as a function of the load of the secondary, which is the target and all of its environment (metal screen, ferrite, battery, etc.):

$$E = [(\,R1 + \frac{M^2\omega^2\ R2}{R2^2 + X2^2}) + j\,(X1\,\frac{M^2\omega^2\ X2}{R2^2 + X2^2})]\,I1 = Zp\ \ I1$$

This equation leads us to the following remarks:

– the load, the target antenna (e.g. that of a mobile phone or a tablet or an NFC key in card emulation mode, etc.) presented to the secondary, provides additional terms in the initial equation (with no load), which tend:

- first, to *increase* (sign "+") the value of the real part R1 of the quantity:

$$R2\,(\frac{M^2\omega^2}{R2^2 + X2^2})$$

- second, to *alter* the value of the reactance perceived by the primary (impedance perceived by the primary) depending on the value (and the positive or negative sign) of X2, which depends on the value of:

$$X2\left(\frac{M^2\omega^2}{R2^2+X2^2}\right)$$

– the value to the secondary of X2 (sooner or later in the form L2ω – 1/C2ω – a circuit whose resonance takes place for $\omega 2o^2 = 1/L2C2$) depends on the value of the working ω. Generally, the value ω2o is:

- equal to ωf, and in this case the value of X2 is null;

- or *ωf is < at the resonance frequency of X1 and ω2o of X2*: the values X1 and X2 are, therefore, capacitive, and X2 contributes an inductance in image form to the primary ("- × - = +"). (Often, for reasons of possible arrangement of the cards or tags in stacks, we set ω2o and the whole of the card at a value higher than ωf);

- *ωf is > at the resonance frequency of X1 and X2*: for reasons of possible arrangement of the cards or tags in stacks, we set ω2o at a value higher than ωf. In this case, at ωf, X2 appears "capacitively", and its value is negative, and thus behaves like an inductance to the primary (with the double "–" sign becoming "+") and thus participates in the detuning of the primary circuit.

8.2.2.2. *Resonance perceived by the primary*

In the condition

$$X2\left(\frac{M^2\omega^2}{R2^2+X2^2}\right) = X1$$

the whole of the reactive term disappears from the equation and the resonance perceived by the primary takes place. In this case, the equivalent generator (E, R1) flows into the load "R2 $M^2\omega^2/(R2^2 + X2^2)$", and the maximum amount of power is transferred.

NOTE.– If, in the application, the quality factor Q2 = L2 ω/R2 of the secondary tuned circuit (the target) is large (>20, for instance), but not too large (<35) to satisfy the quality of the signals prescribed by ISO 14443, so L2 ω > R2 and $(L\omega)^2 = X2^2 >> R2^2$, then this term is equal to:

$$X2\left(\frac{M^2\omega^2}{R2^2+X2^2}\right) = \frac{M^2\omega^2}{X2}$$

Using the above equations to quantify the *loading effect due to the load of the target*, it is then easy to establish the ratio between the values of the currents present in the antenna of the "initiator – base station", non-loaded and loaded, which gives us:

$$\frac{I1}{I1o} = \frac{[\,R1 + \quad + \;\; j\,(X1)}{[\,R1 + \dfrac{M^2\omega^2 \;\; R2}{R2^2 + X2^2} + j\,(X1 - \dfrac{M^2\omega^2 \;\; X2}{R2^2 + X2^2})]} = \frac{(I1)\ \text{loaded}}{(I1_0)\ \text{non-loaded}}$$

This equation obviously depends on the load R2, X2 and on the distance through the mutual M which is a function of the coupling factor k (see Chapter 7), and therefore of the distance "d" between the initiator and target. In any case, all of this tends to decrease the initial value I1o of the no-load current circulating in the initiator, and therefore circulating in the coil of the initiator antenna, and thus to decrease the magnetic field produced and radiated by that antenna, consequently reducing the working distance. It may even reduce the working distance to zero, so that the system can no longer function at all.

This is one of the fundamental reasons for the problems encountered in RFID and NFC.

Now, we simply need to reflect on the content of this equation to deal with the problems, because we have just discreetly introduced the profound *loading effect*.

8.2.3. *Value of R2 in view of the environment*

Often, the most difficult thing is to evaluate the resistive value "ΔR2" equivalent to the contribution of the environment (casing, box, printed circuit board, etc.) to the value of R2:

– for this, without the environment, in air, for a value of R2o (known or easily determinable), a given mutual M (or given k), and thus a given distance and for the desired tuning of the tag, we measure the corresponding value of the current I1o';

– then, in the presence of the desired environment, having taken care to reestablish the tag's initial tuning, and all the rest identical, we measure the new value I1o'' of the current, and obtain a new value R2 = R2o + ΔR2 in the equation, which gives us the ΔR2 specific to the environment and gives an indication of the "weight" of the loading effect.

8.3. Some food for thought

Let us begin by examining the hypothesis (which, in practice, is incomplete and a little false) of the simplest theoretical case – generally that where the users intend to work with X1 = 0 and X2 = 0, consisting of supposing that the primaries and secondaries are tuned on both sides. Certainly, we can always dream.

In practice, the hypothesis that X2 = 0 at the working frequency is only rarely realized (most targets being deliberately shifted toward higher frequencies in order to satisfy their stacking and to easily manage data collisions. Otherwise, the collision management prescribed by ISO 18092 NCF IP1 and ISO 21481 NFC IP 2 would be useless).

The equation of I1/I1o is greatly simplified, and we obtain:

$$\frac{I1}{I1o} = \frac{R1}{\left[R1 + \frac{M^2\omega^2}{R2}\right]} = \frac{(I1)\text{ loaded}}{(I1_0)\text{ non-loaded}}$$

R1 seems to be increased by $M^2\omega^2$/R2serial

with M $= k\sqrt{L1L2}$, or $M^2 = k^2\,L1L2$

$$\frac{I1}{I1o} = \frac{R1}{\left[R1 + \frac{k^2\,L1L2\omega^2}{R2}\right]} = \frac{(I1)\text{ loaded}}{(I1_0)\text{ non-loaded}}$$

or Q2 = Q2serial = L2ω/R2, so therefore

$$\frac{I1}{I1o} = \frac{R1}{\left[R1 + k^2\,L1\,Q2\,\omega\right]} = \frac{(I1)\text{ loaded}}{(I1_0)\text{ non-loaded}}$$

From this equation, we can conclude that, in order for there to be no influence due to the loading effect:

– the value (k^2 L1 Q2 ω) needs to be as low as possible;

– in the worst-case scenario, at all times, with all other things being equal, in order for the ratio between the currents to be as close to 1 as possible, the additional term R1 in the denominator must appear low and constant during the system's operation. With this goal in mind, the value of

(k^2 $L1L2\omega^2/R2serial$) must be constant as a function of the working distance – i.e. R2serial varies proportionally to k^2. However, the parallel view of the secondary circuit enables us to write:

$$Q2 = Q2serial = L2\omega/R2 = Q2parallel = Rp2/L2\omega$$

and expressed as the serial equivalent:

$$R2\ serial \sim= Rp2/Q2^2$$

Returning to the above equation, we divide the top and bottom by R1, obtaining:

$$\frac{I1}{I1o} = \frac{1}{\left[1 + \dfrac{k^2\,Q2\,L1\omega}{R1}\right]} = \frac{(I1)\ \text{loaded}}{(I1_0)\ \text{non-loaded}}$$

and by setting $L1\omega/R1 = Q1$ and $n^2 = k^2Q1Q2$, we find:

$$\frac{\text{loaded}}{\text{non-loaded}} = \frac{I1}{I1o} = \frac{1}{[1 + k^2\,Q1Q2]} = \frac{1}{1 + n^2} \quad \text{“n”} = \text{“coupling index”}$$

This being the case, the approach consists of:

– defining an acceptable value of n_min to define the max. of the ratio I1/I1o;

– and then keeping n_min constant, i.e. with a fixed value of Q1 such that, as a function of the working distance between the initiator and the target, the quality coefficient Q2 decreases while the value of the coupling coefficient increases, so the product k^2Q2 remains constant. In conclusion, as a function of the distance, we need to vary Q2 (and thus Rp2, and hence R2serie) in $1/k^2$… which, in time, leads to the presence of a very particular shunt resistance (see later on in this chapter).

Let us examine another hypothesis (which is a little more accurate) whereby *we suppose only that* $X1 = 0$ *and that* $X2 = x$ at the working frequency ωf, meaning that $\omega 1o = \omega f$, tuning of the base station (which is almost always true), we can write:

$$\frac{I1}{I1o} = \frac{R1}{\left[R1 + \dfrac{M^2\omega^2 \; R2}{R2^2 + X2^2} + j\left(\dfrac{M^2\omega^2 \; X2}{R2^2 + X2^2}\right)\right]}$$

where $M^2 = k^2 \, L1L2$:

$$\frac{I1}{I1o} = \frac{1}{\left[1 + k^2 \, Q1 \dfrac{L2\omega}{(R2^2 + X2^2)} (R2 - j\,X2)\right]}$$

This equation is very realistic, because it merely requires the initiator to be tuned to the working frequency, which (for a host of other reasons) is very often the case. It also takes account of the fact that the target may be tuned to a different frequency from the setup's working frequency, which is also very often the case (because, for reasons of management of the stack of cards, the targets are often shifted slightly higher than the working frequency). Let us expand the equation:

$$\frac{I1}{I1o} = \frac{1}{\left[1 + k^2\,Q1 \dfrac{L2\omega}{R2(1 + X2^2/R2^2)} - j\,k^2 Q1 \dfrac{L2\omega \; X2}{R2 \times R2\,(1 + X2^2/R2^2)}\right]}$$

Let us set "p" = X2/R2 = "tan φ2", in which case:

$$\frac{I1}{I1o} = \frac{1}{\left[\left(1 + \dfrac{k^2\,Q1Q2}{(1 + p^2)}\right) - j\left(k^2 Q1Q2 \dfrac{p}{(1 + p^2)}\right)\right]}$$

$$\frac{I1}{I1o} = \frac{1}{\left[\left(1 + \dfrac{n^2}{(1 + p^2)}\right) - j\left(n^2 \dfrac{p}{(1 + p^2)}\right)\right]}$$

Hence, the crucially important thing is to know the level at which, at ωf, "p = X2/R2" becomes negligible in relation to 1, and therefore we may or may not need to return to the previous section.

8.4. Loading effect

Having mentioned this term so many times, in order to remove all ambiguity and misunderstanding, let us generalize and finally define exactly what the "loading effect" is.

8.4.1. *Definition and comments*

The "loading effect" denotes the physical effect and all of the influences and repercussions (be they *static* or *pseudostatic*) of the load (in all its forms) represented by the general presence of the NFC target, on the initiator's initial (no-load) operation. Physically, the main issue is the effect of the variations in load on the value of the current circulating in the initiator's coil (or target in active mode), between the no-load and loaded states – and therefore on the magnetic field produced – which is where the name comes from.

This "loading effect" is very familiar to professionals in the domain of contactless chip cards and RFID, and it is also very characteristic of NFC applications – particularly of those built around NFC mobile phones, operating in split time in contactless chip card emulation mode, and intending, desiring or needing to be interoperable with existing reader equipment and to conform to the test standard ISO 10376-6 from ISO 14443. At time of writing, some details have been published in the contactless chip card standards, and a little more in the RFID standards ISO 18000-x. However, there is little said about the loading effect in the NFC IP1 and IP2 and NFC Forum standards which are currently available.

By contrast, "load modulation", which we have discussed at length, is a deliberate, mainly *dynamic* effect, which is caused by dynamically modulating the load of the target to the rhythm of the data to be transmitted, to create the retro-modulation signal on the "return channel", in the case of communication between targets and initiators. Load modulation, for its part, is very well detailed in ISO 14443 and 10373-6.

Thus, care must be taken not to confuse the "loading effect" discussed here with "load modulation".

In contactless chip card systems, made of plastic in ID1 format, in conformity with ISO 14443, the problems of loading effects are generally not taken into account, because they are non-existent (simple memory card) or slight (microcontroller card, passport, consuming more power for operation + encryption), and physically, they are practically only loaded by the dynamic "consumption" of the card, and very little by the card's external environment, which is extremely light (a piece of plastic).

This pseudo-ideal case is completely illusory in the case of NFC applications. Indeed, the form factors and loads of NFC devices are very different from those of the chip cards. For example, when a target of a mobile phone in NFC card emulation mode is presented to an initiator, it presents a non-neutral set to the magnetic field radiated by the initiator (metal parts of its casing (iron, copper, etc.), presence of the battery, the integrated circuits, etc., effect of the owner's hand). In addition, there are extremely numerous variants of mobile phone – different models, different makes, different ways and means of presentation of the device to the initiator (screen up, screen down, face side (screen), battery side, flip models, open flip cover, closed flip cover, open/closed sliding models, etc.) – in short, an immense host of possibilities.

Thus, very often, through lack of awareness, the designers of NFC systems or applications have overlooked these effects, and then on the ground, have found themselves faced with serious functional problems in terms of working distance, or occasionally even the lack of any distance at all. We can sum up all of these eventualities by the correct "politico-physico" term "form factor scalability" of the NFC device.

As (unfortunate) examples, we can cite the fundamental and crucial problems of mobile phone applications with an NFC option, operating mainly in card emulation mode is the frequently-envisaged use case of many applications in transport, access control, payment, mPOS, etc. In these applications, we are often in the presence of a system functioning physically in passive mode with alternate bidirectional exchanges (energy + data transmission, followed by the phenomenon of retro-modulation for data return) in magnetic near fields, based essentially on magnetic couplings between tuned circuits.

8.4.2. *Parameters involved in the loading effect*

The main parameters governing the operation of the setup are the values of the mutuals M, the magnetic coupling coefficients k and the coupling indices n which have been discussed at length in the Chapter 7, and which are, themselves, directly functions of the following parameters:

– at a known working distance between the initiator and target antennas:

- the mechanical shape of the coils making up the antennas of the initiator and the target, and thus of their associated electrical inductance values of the coils in antennas L1, L2, their resistances R1, R2 and their quality coefficients Q1, Q2;

- the axial positions of the geometric centers of those coils, in relation to one another;

- the consumption, at all times, of the integrated circuit of the load, on the target antenna (card emulation mode or otherwise) which, itself, depends on:

- the nominal load;

- the nominal consumption of the IC (memory card or µC);

- the bit rate of the communication (106, 212, 424, etc., kbit/s);

- the dynamic variation of the load depending on the application.

This variation of load is mainly dependent on the bit rate and the type of activity of the application. For example, it will be only slight during the initialization procedure (at 106 kb/s), but greater during the phases of "encryption" calculations during the payment phases:

- consumption due to the instantaneous operation of the encryption unit;

- dynamic variation of the communication bit rate;

- the activation of the NFC part depending on the possibilities of the NFC devices;

- dynamic load variation of the load's consumption to form the return link: load modulation/retro-modulation. This is a specific loading effect which, in terms of function, serves for the return communication;

– at constant distance and shape: the immediate environment (magnetic, metallic – copper, aluminum, etc. – ferrite, battery, effect of the hand, etc.) which may have a direct influence on the tuning and detuning and the performances of the magnetic coupling setup;

– the instantaneous distance between the initiator and target: on the one hand, the linear distance between the initiator and target is essentially variable with the nearing and distancing of the hand holding the "NFC device"; and on the other hand, the initiator must be capable of reading numerous types of targets without knowing the values of the field strength which drives them. Therefore, to be safe, the initiator produces a strong field and the targets which are presented, coming from far away and coming ever closer to the initiator, are subjected to increasingly intense fields. Normally, to avoid the harmful effects of these intense variations in fields (and thus voltages induced), on board the target we have an element for regulation and internal protection known as a "shunt regulator", so the variations of impedance as a function of the magnetic field emitted by the initiator (and therefore of the distance) are intended to produce a loading effect appropriate for the situation, without unduly disturbing the rest. This is the case in all well-made chip cards (made of plastic), which are, of course, all in card-emulation mode. This is rarely the case for a mobile phone which, before it is switched on, does not know and cannot guess whether, in the next few moments, it is going to play the role of an initiator or a target. Therefore, the shunt function is very often inexistent or of poor quality, and it is often the mass of its case, electronics and battery which act as a shunt, overwhelming the field.

Figure 8.3 illustrates the loading effect in terms of the main points mentioned above.

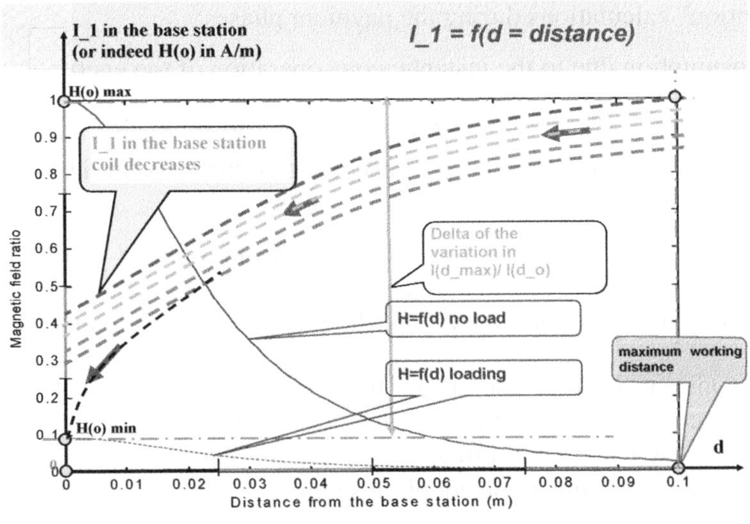

Figure 8.3. *"Loading effect" of the main parameters. For a color version of the figure, see www.iste.co.uk/paret/antenna.zip*

NOTE.– Also, depending on the couplings and/or the loading effect, it occurs that an initiator is capable of communicating with a target at a certain distance, but when the transponder is brought closer to the initiator, the system no longer works.

We will now look at most of these points in turn.

8.4.3. *Variation of the working distance and thus of the coupling*

In HF, at 13.56 MHz, the operating principle is near-field and thus the operation is assured by inductive coupling between the antenna of an NFC device (initiator or target) and another NFC device (again, an initiator or a target), depending on whether it is working as a reader, writer, card emulator or a peer-to-peer terminal.

Depending on the type of NFC device (PC, tag, mobile phone, etc.) and how it works, the coupling values to be achieved are, functionally, nothing like those encountered in standard contactless applications between base stations and conventional chip cards in ID1 format. Indeed, the immediate electromagnetic environment of the antenna of a contactless chip card is generally very "light" (an antenna with a few spires sandwiched by a plastic casing). An NFC device such as, say, a mobile phone, is completely different, and its environment (metal casing, battery, numerous integrated circuits, screen, etc.) plays a significant role in the degradation of the magnetic coupling setup.

8.4.4. *Magnetic coupling and its consequences*

8.4.4.1. *Form factors of the antennas*

One of the main problems of NFC lies in the definition of NFC devices, initiators and targets, in terms of their physical appearance, mechanical shapes, large, small, simply known as "form factors", and of course of the antennas, their shapes and relative magnetic couplings, arising:

– first, on the applicative fields (mobile telephony, consumer electronics, etc.);

– second, on the application in the same applicative field (e.g. for different models of mobile in the family of mobile phones, flip devices, sliding phones, etc.)

8.4.4.2. *Influences of the antennas' near environments and the form factors*

Once the antenna's "form factor" is known (often defined by the mechanical design of the NFC device), for simple reasons of conformity tests, we also need to estimate and verify its performances in the working environment where the NFC device needs to operate. Of course, that environment itself can vary during application (for instance, when the user opens the flip cover of a mobile phone while using it to enter a code in "touch and confirm"). Surprising? Take a simple flip phone for example. Regardless of the antenna's physical position (in the flip cover or in the actual body of the device, on the battery side or not), we need to ensure that it conforms in all positions of the flip cover (notably whether it is open or closed).

8.4.4.2.1. Influence of the environment

The environment is a function of all sorts of parameters – magnetic, metallic, primary effects, electromagnetic effects, etc. Once this long list has been evaluated for the desired application, we then need to model these effects, simulate them and measure them to ensure the applications and all their derivatives and variabilities during all phases of use, depending on the environments.

8.4.5. ***Performances required by the initiator: loading effect on the value of the remote power supply to the target***

As soon as the antenna of an NFC initiator is loaded by an NFC target (or future initiator), because of inductive coupling, the target causes a loading effect. Depending on the mechanical and/or electronic implementations and the initiators and targets/tags, this effect may significantly alter (in fact, reduce) the current circulating initially (no load) in the initiator antenna, and hence decrease the value of the magnetic field radiated "to the load" by the initiator (see the real top of the chapter). Sometimes, the field may no longer be strong enough to properly remote-power the target. This problem occurs primarily when the coupling coefficient between the initiator antenna and the target antenna is too strong, leading to the initiator no longer being able to supply the energy/power necessary to correctly power the tag.

8.4.6. *Quality of the emitted magnetic field*

The purpose of the following sections is to analyze these effects and their consequences in detail, with the case study of the applicational framework of NFC standards IP1 and IP2 in which the initiator, so as to conform to the ISO standard, is just capable, without a load, of providing the minimum magnetic field of 1.5 A/m to a target (often in batteryless card-emulation mode) which, as it absorbs energy, significantly decreases the value of that field as we approach the target, which actually has the effect of causing gaps in communication.

8.4.6.1. *Detailed study of the loading effect*

In order to examine the consequences of the loading effect, it is necessary to look at the value of the current circulating in the antenna of the NFC initiator during the presence of the target, and the phase of transmission as a function of the distance between the initiator and target, because the magnetic field generated by the initiator is directly proportional to the current running through it.

As previously seen, a model of the coupling between the initiator and target (see Figure 8.4) can be used to calculate "the ratio between the current in the initiator antenna in the absence of a tag and the current in the initiator antenna in the presence of the tag".

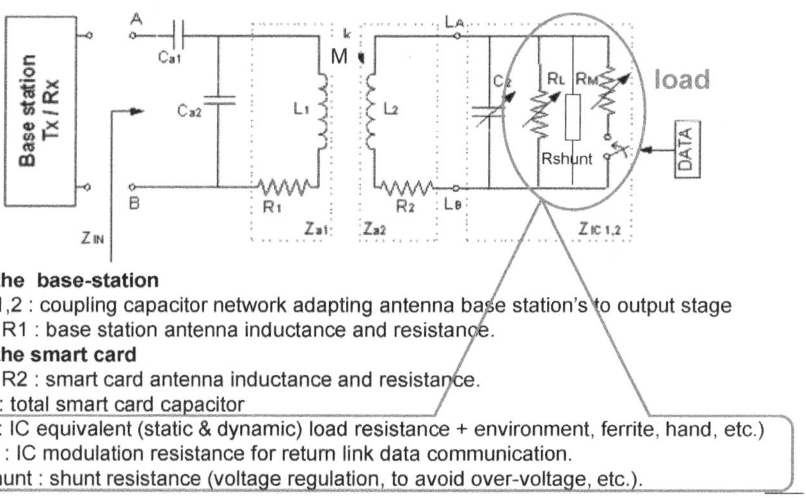

Figure 8.4. *Model of simulation of coupling between initiator and target*

Furthermore, as we will show at the end of this chapter, another point is to optimize the law of variation of the resistance of the shunt regulator of the target, as a function of the field received (and therefore of the distance) in order to minimize the loading effect (see section 8.4.8).

8.4.6.2. *Ratio between the currents in the non-loaded/loaded initiator*

The equivalent diagram in the Figure 8.4 demonstrates most of the components implemented concretely, where:

– at the level of the initiator:

$C_{a1.2}$ capacitive impedance-matching network of the initiator antenna;

L_1, r_1 inductance and resistance of the initiator antenna.

– at the level of the target:

L_2, R_2 inductance and resistance of the target antenna;

C_2 total capacity of the NFC target (equivalent input capacity of the integrated circuit + parasitic capacity of the antenna + capacity of encapsulation of the integrated circuit);

R_L equivalent resistance of the total load (integrated circuit + equivalent environment). In addition, this resistance also depends on the mode of operation of the integrated circuit;

R_M resistance of load modulation for retrotransfer of data;

R_{shunt} shunt resistance to avoid oversupply of the NFC target which we are seeking to optimize;

R_{target} or R_{tag} resistance equivalent to the setup $R_L//R_M//R_{shunt}$;

M mutual inductance between the antennas.

with $M = k\sqrt{L_1 L_2}$

and $k = \mu \dfrac{r^2}{2\left(r^2 + d^2\right)^{3/2}} n_1 n_2 s_2 \dfrac{1}{\sqrt{L_1 L_2}}$

where

μ = μ_0 μ_r	= 4π10⁻⁷ μ_r, apparent magnetic permeability;
r:	the radius of the initiator antenna;
N1:	number of turns of the initiator antenna;
N2:	number of turns of the target antenna;
s2:	surface of a spire of the target antenna;
d:	distance between the two antennas.

With the equations demonstrated at the beginning of this chapter, this diagram enables us to quantify the ratio between the currents circulating in the initiator antenna when it is not loaded and when it is. Figure 8.5 clearly shows the points of action of the different parameters.

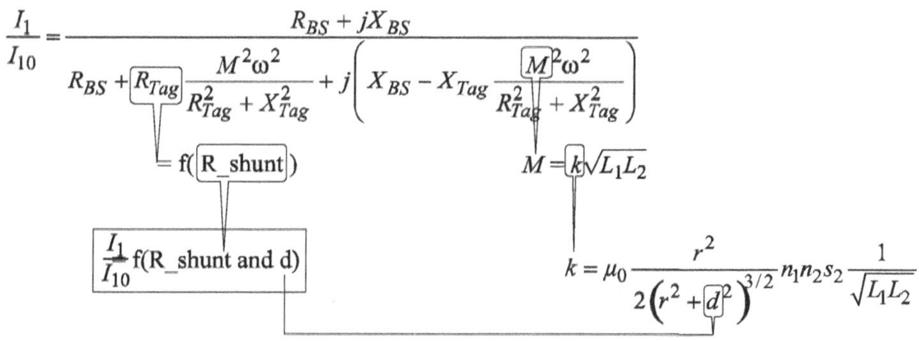

Figure 8.5. *Parameters affecting the ratio between the currents circulating in the initiator antenna without a load and with one*

8.4.6.3. *Consequences for the magnetic field produced*

The different loading effects lead to a drop in current I1 in the initiator. However, the magnetic field radiated by its antenna is proportional to that current. If the current falls too much, the magnetic field radiated by the initiator may fall below the minimum value given by the NFC standard IP1, which is 1.5 A/m and the system no longer works, as Figure 8.6 shows.

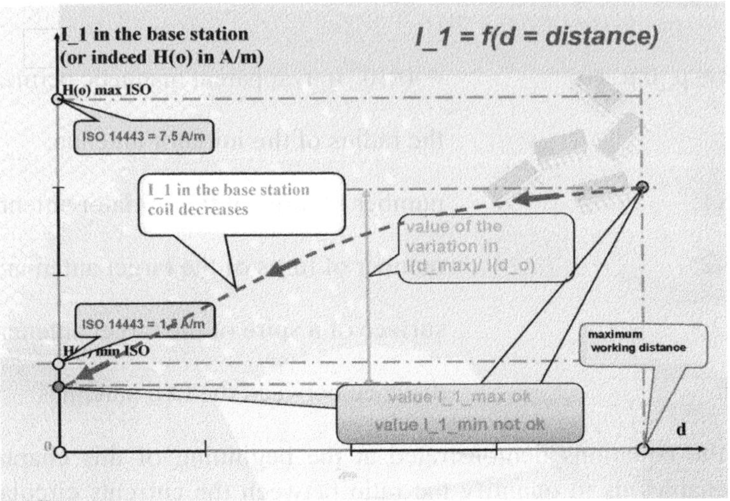

Figure 8.6. *Solution without shunt with an insufficient final value of the field. For a color version of the figure, see www.iste.co.uk/paret/antenna.zip*

It is not desirable to increase the current in the initiator because, when the initiator is not loaded (the target is a long way away), there is then a risk of producing too powerful a magnetic field (stronger than 7.5 A/m) – see Figure 8.7.

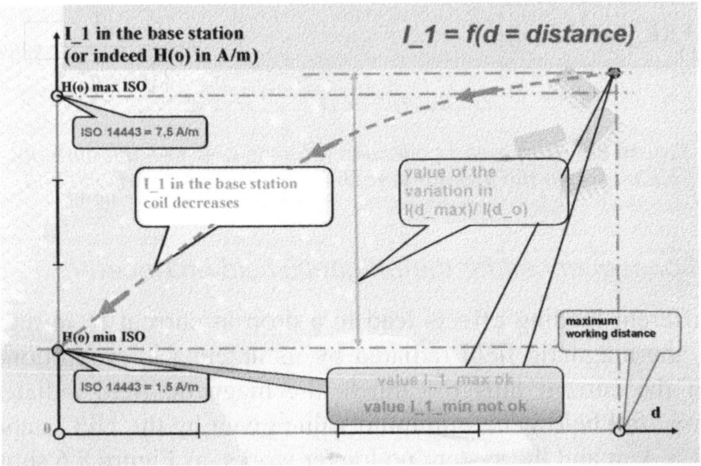

Figure 8.7. *Solution without shunt with too high an initial value of the field. For a color version of the figure, see www.iste.co.uk/paret/antenna.zip*

As we will see a little later on, the solution is to have a shunt to bring the field back within the range allowed by the standard (1.5 min–7.5 max. A/m – Figure 8.8).

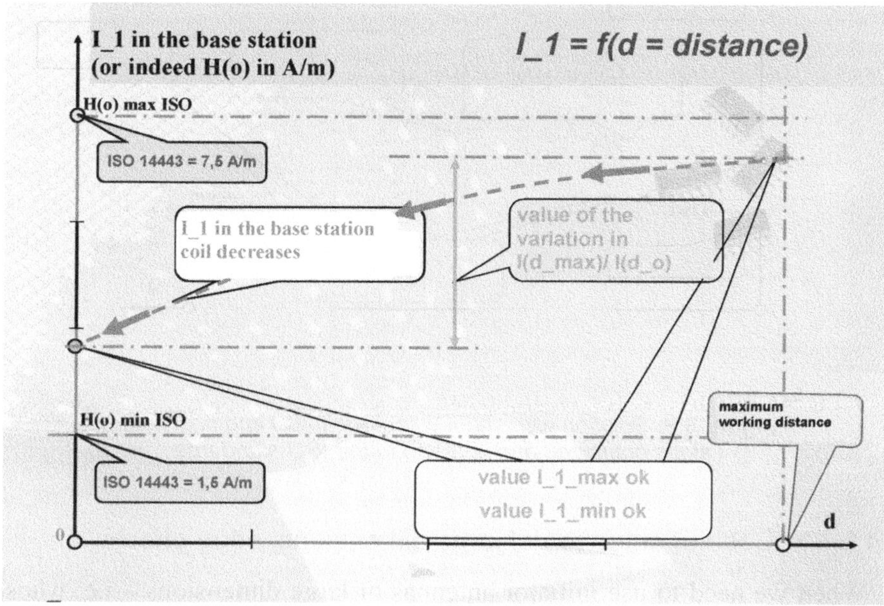

Figure 8.8. *Solution with shunt with initial and final values of fields, conforming to the ISO standards*

The final thing we need to do is to optimize the law of variation of the shunt resistance in order to minimize the variations of the magnetic field radiated as a function of the distance between the initiator and target (Figure 8.9).

We seek to optimize the law of variation of the shunt so that the magnetic field respects the minimum and maximum values stipulated by the standard.

8.4.6.4. *Consequences of the size of the initiator antenna*

The next this section draws on a document from AFIMB (French agency for multimodal information and smart-ticketing), provided to the author by Jean-Paul Caruana at GemAlto.

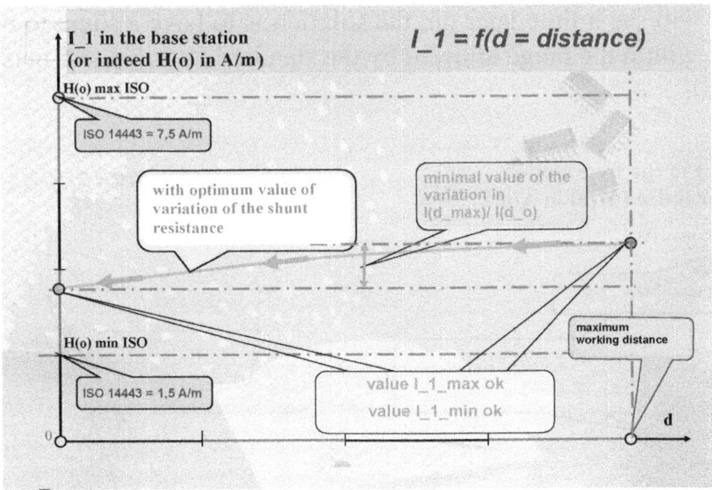

Figure 8.9. *Solution with shunt where the initial and final field values optimized and conform to the ISO standards*

8.4.6.4.1. Case of antennas of large dimensions: edge effects

When we need to use initiator antennas of large dimensions – i.e. whose equivalent diameter is around 10–15 cm – to ensure a minimum field of 1.5 A/m at its center, then at a distance of 0 cm, the initiator must produce a field of at least 0.5–1 A/m more than that minimum at the edge of the antenna. Indeed, as Figure 8.10 illustrates, all around the spires, the intensity of the field is substantially greater than at the center (when we have a field of 1.5 A/m at a distance of 5 cm from the plane of an antenna 14 cm in diameter, the field intensity can easily be twice that value at the spires).

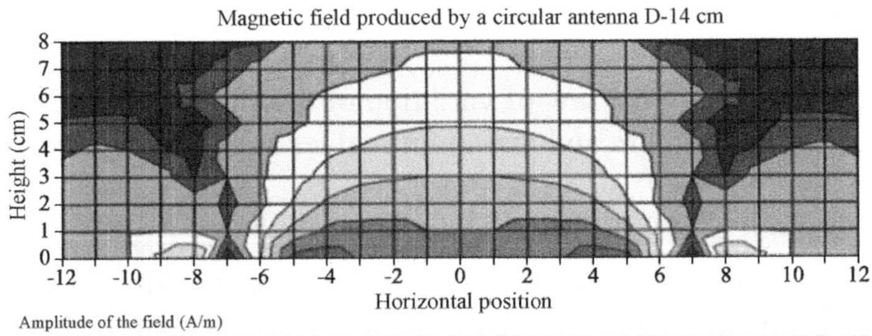

Figure 8.10. *Case of large antennas – edge effects*

8.4.6.4.2. Case of small antennas: "loading effect"

When the initiator antenna is small – i.e. its diameter is less than ~10 cm – the risk of strong coupling with the target means that we must produce a field at the surface of the antenna which is stronger than the minimum desired field (even, for instance, that specified by ISO/IEC 14443 … for the class of antenna in question). Indeed, because of the presence of the target, the "loading effect" must be withstood by the initiator. Generally, the minimum loading effect is around 0.3 A/m. For example, an initiator with an antenna whose surface is smaller than 2 times that of an ID1 card must produce a minimum field of 1.8 A/m, instead of 1.5 A/m, in its working volume. An example with an antenna 6.5 cm in diameter is presented in Figure 8.11.

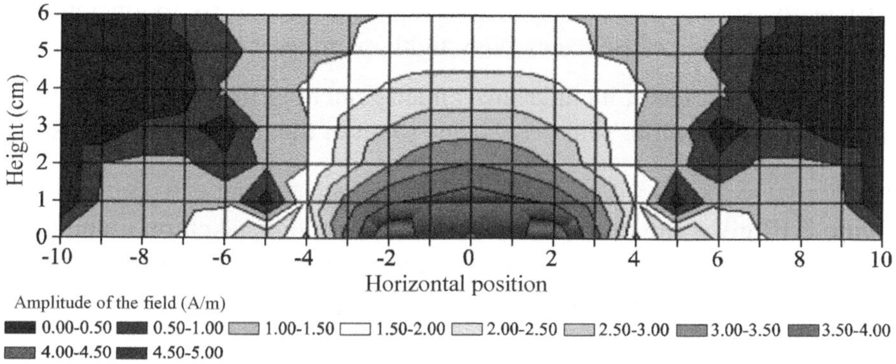

Figure 8.11. *Case of small antennas – the loading effect*

Obviously, reducing the size of the initiator antenna is a step in the right direction because, with the tag's "s2" remaining the same, the value of "r" decreases and that of "k" increases, but the loading effect is more marked.

8.4.6.4.3. Conclusion

Once again, in the present case, this is still insufficient because, owing to the fact that "s2" is small, we would need to greatly reduce the size of the initiator antenna in order to obtain a viable solution, and often, the new antenna size would be incompatible with the ergonomics desired for the solution.

8.4.6.5. *Extremely important conclusions*

When developing an NFC product, above all we need to be careful not to amalgamate and confuse standard "chip cards" (made of plastic, format ID1, etc.), with an NFC device operating in card emulation mode, whatever that device is (a mobile phone, an e-wallet, etc.), with its form factor, favorable or unfavorable effect on the magnetic fields, etc.

The solution?

– First, design an initiator antenna which is sufficiently larger than that of the target to decrease the coupling, even when the two are very close together;

– second, have a no-load magnetic field for the initiator which is sufficiently strong that the system will never be forced to operate at a distance that would cause excessively tight coupling of the setup;

– third, as a reserve, include smart management of the loading effects and variation of the specific shunt resistance shunt of the targets (see later on).

CONCRETE EXAMPLES.– In order to help visualize what is being said here, Figure 8.12 gives the examples of real measurements taken on a number of commercial devices (anonymized here), and gives a concrete illustration of the problems that need to be dealt with.

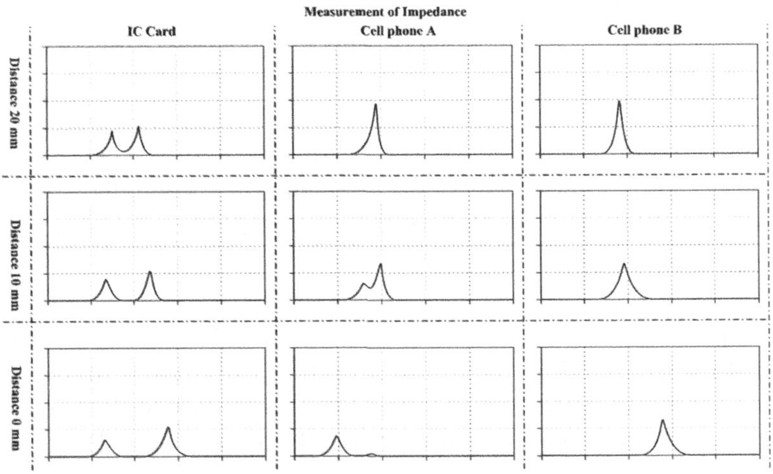

Figure 8.12. *Examples of real measurements taken on different commercial devices*

8.4.7. *Examples of coupling coefficients and loading effects*

For didactic purposes, let us begin by looking at the example of an *initiator* whose *antenna has deliberately been designed to be small* (NFC initiators such as a mobile reader, portable reader or portable banking terminal, etc.), where the characteristics of the antenna are (class 5, ISO):

- length a = 40 mm;
- width b = 20 mm;
- track width l1 = 1 mm;
- number of turns N1 = 1 so, after calculations, L1 = 664.5 nH.

We will now examine two examples of specific applications corresponding to two types of targets, presented to the same initiator.

8.4.7.1. *Example 1: target in classic ID1 format of an ISO card*

- length a = 72 mm;
- width b = 42 mm;
- track width l2 = 0.15 mm;
- number of turns N2 = 4 so, after calculations, L2 = 3.17μH.

Now that we know the values of L1 and L2 and the respective number of turns of the antennas, and supposing that the axes of the antennas are colinear, we can calculate the variations of the coupling coefficient by using the same formula that we have already seen many times (Figure 8.13):

$$k = \mu \frac{r^2}{2\left(r^2 + d^2\right)^{3/2}} n_1 n_2 s_2 \frac{1}{\sqrt{L_1 L_2}}$$

When the target is placed at the center of the initiator antenna (at no distance at all – d = 0), the coupling coefficient is maximal, and equal to:

k_0 = 0.16 = 16%

with Q1_initiator = 15 (including, of course, the source impedance of the generator).

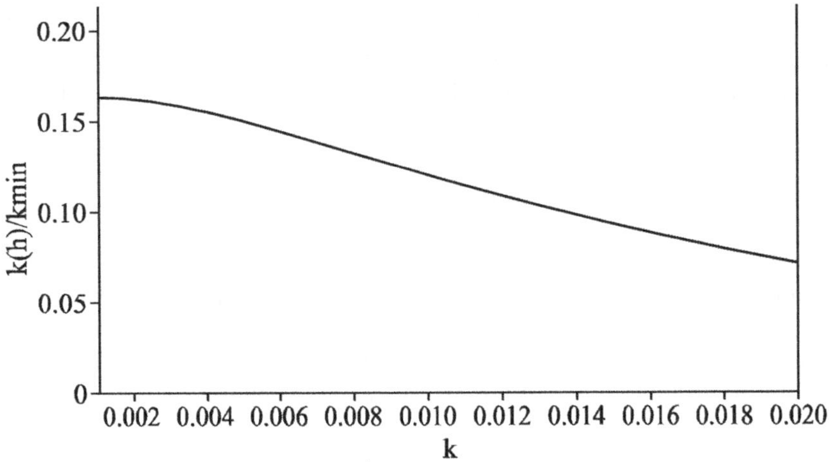

Figure 8.13. *Variations of the coupling coefficient as a function of the distance*

This gives us a ratio of loaded to non-loaded current I/Io of 0.44 = 44%

The current circulating initially in the no-load initiator antenna is, therefore, reduced when loaded by approximately 56%, meaning that, in order for the loaded initiator to deliver a minimum field H of 1.5 A/m, in order to remote-power a target conforming to the ISO (just to standard), then in a no-load situation (where there is no target within the range of operation), the initiator must deliver a 3.4 A/m field.

As this last value is less than the maximum authorized value of 7.5 A/m, the reader thus constituted will always be "ISO 14443-2 and ISO 10 373-6 compliant".

8.4.7.2. *Example 2: target in format with half size of the ISO card*

Still using the same initiator, with its small antenna, let us now present a new *target whose antenna is smaller (by half)*, and repeat the calculations with the following characteristics (see Figure 8.14):

– length a = 70 mm;

– width b = 22 mm;

– track width I1 = 0.15 mm;

– number of turns N1 = 4 so, after calculations, L1 = 2.32μH.

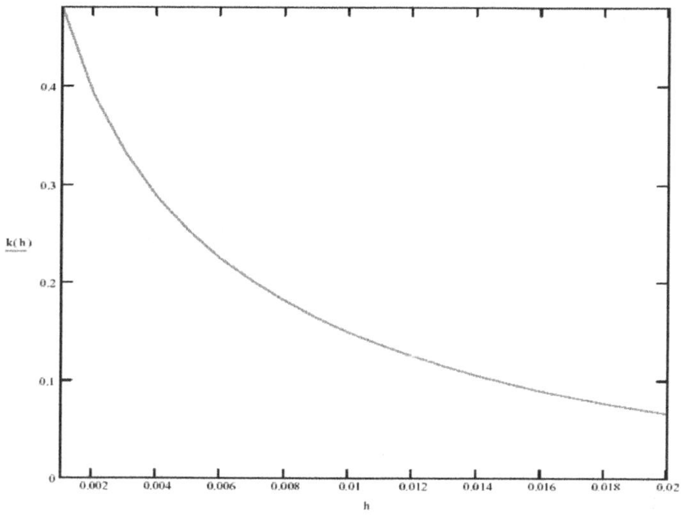

Figure 8.14. *Variations of the coupling coefficient as a function of the distance*

When that target is placed at the center of the antenna of the same initiator (at distance d = 0), the coupling coefficient is now equal to:

k_0 = 0.5 = 50%

where Q1_initiator = 15 (including, of course, the source impedance from the generator).

This gives us a ratio of loaded to non-loaded current I/Io of 0.052 = 5.2%.

The current initially powering the "non-loaded" antenna of the initiator is, therefore, reduced by approximately 94%. This means that, in order for the loaded initiator to deliver a minimum field H of 1.5 A/m, in order to remote-power an ISO-compliant target, then in a no-load situation (where there is no target in range), the initiator must deliver a 29 A/m magnetic field. As this value is much higher than the maximum authorized value of 7.5 A/m (or more for class 5), we can state that it will not be possible to create an "ISO 14443-2 or ISO 18092 and ISO 10373-6 compliant" initiator to remote-power that target.

8.4.7.3. *Further information*

In addition to the above, the curves presented in Figure 8.15 offer an explicit comparison of the values of the coupling coefficients "k" between three NFC solutions often present on the market:

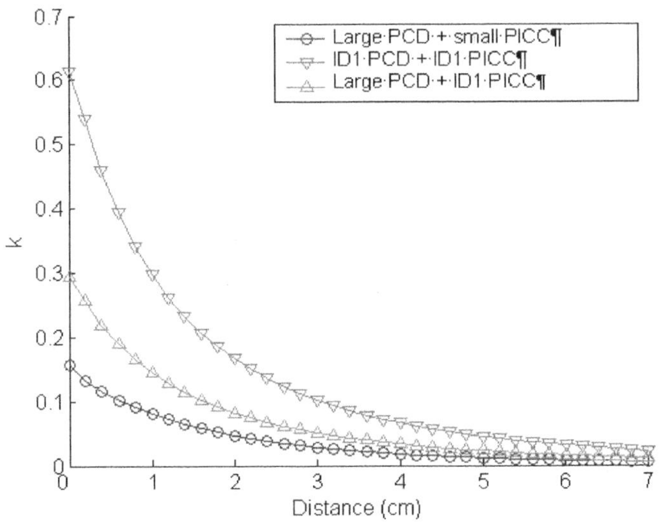

Figure 8.15. *Comparison of the variations in coupling coefficient as a function of the distance. For a color version of the figure, see www.iste.co.uk/paret/antenna.zip*

1) Large initiator antenna (format of the ISO's test tool) with a conventional target in ID1 ISO format (curve marked with triangles): k_0 = 0.3.

EXAMPLE.– Transport validator, with a contactless card (such as the SmartLink system used on the PATH transit system in New York).

2) Large initiator antenna (same format as the ISO's test array) with a smaller target, in a format half that of the ID1 of the ISO. With the same initiator antenna, when the target is smaller, the coupling is reduced (curve marked with circles): k_0 = 0.16.

EXAMPLE.– Public transport validator (such as that used by the RATP in Paris) with a transport ticket in the form of an NFC USB key.

3) Antenna initiator in the same ID1 format as the target. As the format of the initiator and target is the same, the coupling is better (curve marked with upside-down triangles): k_0 = 0.6.

EXAMPLE.– Portable payment reader with a mobile phone application (PoS and mPOS), well designed in terms of coupling, but less so in terms of the loading effect (see immediate next sections).

8.4.7.4. *Almost concrete examples of applications*

Let us give some orders of magnitude for common applications. With this aim in mind, look again at the case of the second target (k_0 = 0.5) above, and suppose that, with no load, the magnetic field produced by the initiator is:

– first, *ideally* uniform at 0–5 cm (often, this ideal uniformity is not delivered);

– second, set at 2.5 A/m by the designer (who believes she/he has already looked at the bigger picture and adjusted the overall operational consumption of his/her NFC device);

– finally, the quality coefficient Q2 of the target remains constant as a function of the distance (which is not normally the case in the real world, because of the presence of a shunt regulator in the circuit of the target … if it is well designed).

D	Q1	Q2	Coupling			Current		H_0	Kt
						Loaded			
			k coefficient	n index	appellation	I1/I1o	reduction	At center	
(cm)				k √Q1Q2		$1/(1+n^2)$	in %	(A/m)	$n/(1+n^2)$
5	–	–	0.000	–	–	1	0	2.5	–
3.2	15	15	0.03	0.45	loose	0.85	15	2.12	0.4
1.4	15	15	0.1	1.5	critical +	0.33	67	0.82	0.47
0	15	15	0.5	7.5	very tight	0.1	90	0.25	0.75

Table 8.1. *With H being constant and Q2 of the target also constant (without shunt)*

Still using the same hypotheses (which do not perfectly reflect reality), in the case where the quality coefficient Q2 decreases with distance (with a shunt regulation present in the target), an improvement could have been attained.

D	Q1	Q2	Coupling			Current		H_0	Kt
						Loaded			
			k coefficient	N index	appellation	I1/I1o	reduction	At center	
(cm)				$\frac{k}{\sqrt{Q1Q2}}$		$1/(1+n^2)$	in %	(A/m)	$n/(1+n^2)$
5	–	–	0.000	–	–	1	0	2.5	–
3.2	15	15	0.03	0.45	loose	0.85	15	2.12	0.4
1.4	15	2	0.1	0.55	loose	0.65	45	1.62	
0	15	0.1	0.5	0.61	critical	0.35	65	0.87	

Table 8.2. *With H being constant and variable Q2 of the target (with shunt)*

> Consider a given initiator, whose no-load field production is between H_0 and H_5 cm, which is far greater than the minimum set by the standard (2.5 A/m for 1.5 A/m) and a specific target (of a given form factor), whose own minimum activation field is 1.5 A/m (which is just compliant to the ISO standard). We could reasonably hope to operate in the green zone in the tables where, in spite of the loading effect due to the target, the initiator manages to produce, at its center, a field which is stronger than the minimum value stipulated by the standard. What is certain is that the target will not function in the red zone, although everyone believes that we are as close as possible to the initiator, with a field of 2.5 A/m … which is totally fallacious, because it only exhibits values below 1 A/m, even reaching as low as 0.25 A/m.
> Amateurs, beware!

Box 8.1. *Important conclusions for all NFC device developers to bear in mind*

8.4.7.5. *More than concrete examples of applications*

In order to stick as closely as possible to reality and offer a more than concrete example, let us look again at the last case above, equipped with a shunt regulator (see below), and still suppose that in the no-load situation, the field H_0 produced by the initiator antenna is set at 2.5 A/m by the designer (who believes she/he has already looked at the bigger picture and adjusted the overall operational consumption of his/her NFC device), but now:

– to begin with, of course, the magnetic field H_d is not uniform, and decreases as a function of the distance, in accordance with the Biot–Savart law;

– in addition, the quality coefficient Q2 also falls with increasing distance.

Returning to our tables, let us fill in Table 8.3.

	No load						Loaded				
D	H_d	Q1	Q2	Coupling Between the initiator and the target			Current in the initiator antenna		H_0 at the center	H_d at distance	Kt
				k coef.	n index	appellation	I1/I1o				
Cm	A/m				k $\sqrt{Q1Q2}$		$1/(1+n^2)$	in %	A/m	A/m	$n/(1+n^2)$
5	1.5	15	15	0.000	–	–	1	0	2.5	1.5	–
3.2	1.75	15	15	0.03	0.45	loose	0.85	15	2.125	1.487	0.4
1.4	2	15	2	0.1	0.55	loose	0.65	45	1.625	1.3	
0	2.5	15	0.1	0.5	0.61	critical	0.35	65	0.87	0.87	

Table 8.3. *With decreasing H and variable Q2 of the target (with shunt)*

This means that the target will no longer operate at 3.2 cm, because it only receives a weaker field.

In conclusion, in this example:

– Despite the no-load value H_0 = 2.5 A/m, the target operates at a far distance, but not at near ones.

– We would have needed a no-load field H_0 = ~5 A/m in order for it to function at d_0.

– It is best to produce an intense field at H_0, thus giving an above-standard value at distance and present the target at a distance, because this way, there is a lesser loading effect. This is the recommendation made by EMVCo.

8.4.8. *"Shunt" circuit in NFC*

It is here that we begin to be able to distinguish good targets, good tags, etc. from the not-so-good ones.

Typically, an NFC target ought (but only ought) to have an electronic element serving as an internal (parallel) "shunt", whose role is to eliminate and/or regulate the incident voltage/power arriving at the target as a function of the distance (far = weak, near = strong signal) – see Figure 8.16.

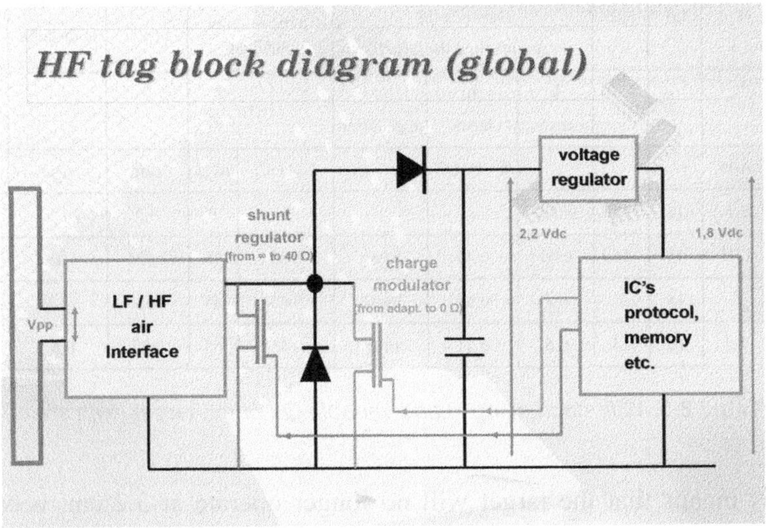

Figure 8.16. *Diagram of a circuit with shunt*

This being said, there is nothing to dictate how the variation of the resistance of that shunt must evolve with changing distance, and have an effect on Q2. If it varies too slowly, it will not produce any effect to compensate the loading effect; on the other hand, if it varies too quickly and too much, it will increase the loading effect and interfere with the operation.

Thus, the law of variation of the shunt resistance needs to be very specific, sagely and finely calculated and tailored to the device. The (lengthy) calculations by which this law is obtained go beyond the scope of this book (but are given and demonstrated in "RFID at Ultra and Super High Frequencies: Theory and Application" by Dominique Paret, published by John Wiley). The main results are recapped below.

8.4.8.1. *Calculation of R_sh(d)*

The law of variation of R_sh with distance = f(d) is written as follows (see [PAR 12]):

$$R\_sh(d) = \frac{(B \times L2\,\omega \times R\_ic)}{\left[\dfrac{(Kte^2\ Q1\ R\_ic)}{(r^2 + d^2)^{\wedge}3} - (B \times L2\,\omega)\right]} = f(d)$$

This law of variation is not easy to respect, and it is the extent to which it is respected which determines the quality of the anti-loading effect. Thus, to do it right, the tag needs to measure the analog value of the field H as a function of the distance, convert that value into digital and consult a table giving the appropriate values of the R_shunt to obtain the optimum value that yields the minimal loading effect (Figure 8.17).

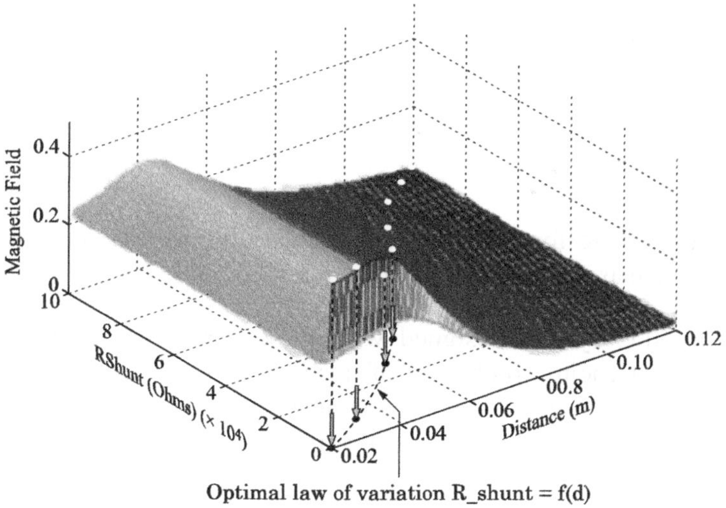

Figure 8.17. *Law of variation of R_shunt yielding the minimal loading effect*

8.4.8.2. *Example of variation of R_sh(d)*

Here is a concrete example.

We wish for "I1, with loading" to always correspond to guarantee a minimum magnetic field of 1.5 A/m regardless of the working distance, and for the value of the ratio I1/I1o as a function of that distance not to vary by more than 90% of the "no-load value I1o". The no-load magnetic field, therefore, must be a field of 1.66 A/m which is always compliant with ISO 14443:

Thus,

$$I1/I1o = A$$

$$A = 1/(1 + n^2) = 0.9$$

which gives us

$$n^2 = (1 - A)/A = B = 0.111$$

so that

$$n = 0.33$$

The system envisaged is as follows (N.B. the values are not chosen at random):

Initiator		Target	
L1	= 500 nH	L2	= 4 μH
N1	= 4	N2	= 4
Q1	= 15	s2	= 40 cm²
r	= 4 cm	R_ic	= 10 k ohm

Using the above formula, we can calculate the necessary values of the shunt resistance as a function of the distance (see Table 8.4).

Is your NFC system designed with these parameters borne in mind? Are you experiencing problems of loading effect when your target is presented to your initiator? Even on contact, does the system work 50% of the time? Yes? You are now fully informed. You now have all the tools at your disposal.

	Parameters	Values					Units
Hypotheses	I1/I1o = A	0.9					%
	B = n²	0.11					
	n	0.33					%
Constants	μ	1.256E-06					
	ω	85156800					rd/s
Distances	d	18	15	10	5	0	mm
Calculations	r²	0.0016					m²
	Kte	4.5472E-05					
	Kte²	2.0677E-09					
	Kte² Q1 R_ic	3.10E-04					
	B L2 ω	37.85					
	B L2 ω R_ic	378474.67					
	R_sh(d)	113164.19	12634.07	2352.87	363.58	5.00	ohm

Table 8.4. *Example of the calculation of the value of the shunt resistance*

8.4.8.3. *In summary, in the presence of a shunt (Table 8.5)*

All these remarks enable us to define a zone/volume of good/healthy operation (SOAR, for safe operating area) for the initiator–target setup (without prejudging the possibilities for demodulation).

8.4.8.4. *Solutions and conclusions*

Hopefully, readers have now gained a fairly healthy understanding of the theory and consequences of the loading effect, which an NFC system will always have to face, because of the different form factors of the NFC devices that will be presented to another NFC device.

This concludes this long chapter, devoted to the loading effect. While the length is regrettable, our long experience in the field has shown that, apart from the typical software bugs attributable to diagonal reading of the specifications, the most difficult operational problems are those primarily due to the form factors, loading effects and couplings.

The target is…			
Far from the initiator	At the theoretical threshold for its activation H_threshold	In the middle of the magnetic field	(Very) close to the initiator antenna
k is very low Q2 is high The integrated circuit of the target is not yet working	k is low Q2 is high The integrated circuit of the target is beginning to work, but the regulator has not yet been activated	k is a little higher Q2 is: a) either the integrated circuit of the target has begun to work and its regulator is not operating – Q2 remains constant b) or the integrated circuit is working and the regulator is beginning to act – Q2 starts to decrease	k increases and becomes significant Q2 is: a) either the integrated circuit of the target is working and its regulator is not operating (or indeed it does not have one) – Q2 remains constant b) or the integrated circuit is working and the regulator is doing its job – Q2 has sharply decreased, and tends toward zero
The influence of the load is slight, and the initiator is unaffected	Because of the low value of k, in spite of the high Q, the product k Q is low The influence of the load is slight	depending on a) or b) the product "k Q2" a) increases (k increases, Q remains unchanged), so the load is beginning to have an influence b) remains constant (k increases and Q2 decreases), so the influence of the load remains slight	depending on a) or b) the product "k Q2" a) has increased sharply (k has increased greatly, and Q2 does not decrease), so the influence of the load is strong b) remains constant, or decreases (k increases and Q2 decreases sharply), so the influence of the load remains slight

Table 8.5. *Summary, in the presence of a shunt*

8.5. Appendix: how do we approach an NFC project?

<table>
<tr><td></td><td>Target</td><td>Ergonomics</td><td colspan="2">Initiator</td><td></td></tr>
<tr><td></td><td></td><td></td><td>Antenna</td><td>Integrated circuit</td><td></td></tr>
<tr><td rowspan="7">1 ↓</td><td>Choice of catalog of the target used</td><td></td><td></td><td></td><td></td></tr>
<tr><td>Type of integrated circuit used in the target</td><td></td><td></td><td></td><td></td></tr>
<tr><td>V_threshold of the integrated circuit used</td><td></td><td></td><td></td><td></td></tr>
<tr><td>Number of spires and equivalent surface of the target antenna</td><td></td><td></td><td></td><td></td></tr>
<tr><td>Calculation of the H_threshold of the target</td><td></td><td></td><td></td><td></td></tr>
<tr><td colspan="2">Desired working distance</td><td></td><td></td><td></td></tr>
<tr><td colspan="2">H_0_base_station necessary = NI / 2 r_equi_bs</td><td></td><td></td><td></td></tr>
<tr><td rowspan="2">2 ↓</td><td></td><td>Max possible bulk of the initiator antenna</td><td></td><td></td><td></td></tr>
<tr><td></td><td>The S_ant_bs gives r_equi_max</td><td></td><td></td><td></td></tr>
<tr><td rowspan="3">3 ↓</td><td></td><td></td><td>Hence the product NI for H_0</td><td></td><td></td></tr>
<tr><td></td><td></td><td>Choice of distribution between N and I</td><td></td><td></td></tr>
<tr><td></td><td></td><td>N and r_equi_max gives the possible value of L</td><td></td><td></td></tr>
<tr><td></td><td></td><td></td><td>L and Q_ of the application gives Rs to be satisfied →</td><td>Hence the calculations of the EMC matching circuits and T-circuit filters ↑</td><td>5 ↑</td></tr>
<tr><td></td><td></td><td></td><td colspan="2">Rs and R_int given Ils fP_max</td><td rowspan="6">4 ↑</td></tr>
<tr><td></td><td></td><td></td><td></td><td>Q_appl necessary</td></tr>
<tr><td></td><td></td><td></td><td></td><td>Thus, P_max</td></tr>
<tr><td></td><td></td><td></td><td></td><td>R_int</td></tr>
<tr><td></td><td></td><td></td><td></td><td>V_out_pp</td></tr>
<tr><td></td><td></td><td></td><td></td><td>Type of integrated circuit used in the initiator</td></tr>
</table>

Table 8.6. *How to approach an NFC project*

Conclusion

Numerous times in this book, we have mentioned "near" and "far" fields and the "near-field communication (NFC)" in its generic sense – i.e. in the physical sense of near-field communication, independently of the value of the working frequency used for the system. As we will see in this conclusion, there were profound reasons for this.

Indeed, under the NFC umbrella, far too many people (users, journalists, speakers, etc.) accidentally or deliberately confuse systems operating at 13.56 MHz with its use in mobile telephony, or with contactless chip-card emulation or substitution. Alternatively, they create an amalgamation which serves their own purposes, by limiting the scope of the term "NFC". However, there are many other industrial views of NFC.

Thus, if readers want to shed their blinkers about "NFC high frequency (HF) for mobile telephony", here are a few useful suggestions.

Technical technological future of NFC "as it is", or almost

Very high bit rates

Section 2 of the contactless chip card standard ISO 14443, in its current edition (2015), specifies a basic/initial bit rate of 106 kbps, with high bit rate (HBR) up to 848 kbps, and also supports two versions of very high bit rates – VHBRs, extending communication up to 27 Mbps. The specifications of NFC IP1 and IP2 are limited to 106, 212 and 424 kbps. The bit rates are changing and, in time, if the ISO standards on NFC follow those on contactless chip cards, the bandwidths and quality coefficients Q of the antennas will also change.

Active load modulation – ALM

Very frequently, throughout this book, we have mentioned active load modulation (ALM). Indeed, in 2015, for battery-assisted applications (smartphones, watches, bracelets, wearable technology, etc.), the principle of retro-modulation from the target to the initiator by passive load modulation (PLM) is in the process of being supplanted by ALM. In these particular applications, ALM is coming on a pace for two main reasons: first, to combat the significant loading effects experienced by NFC devices (because of cases, batteries, etc.), and second, to reduce the dimensions of the antennas (μSD cards, μSIM cards, bracelets, etc.) and the associated coupling values.

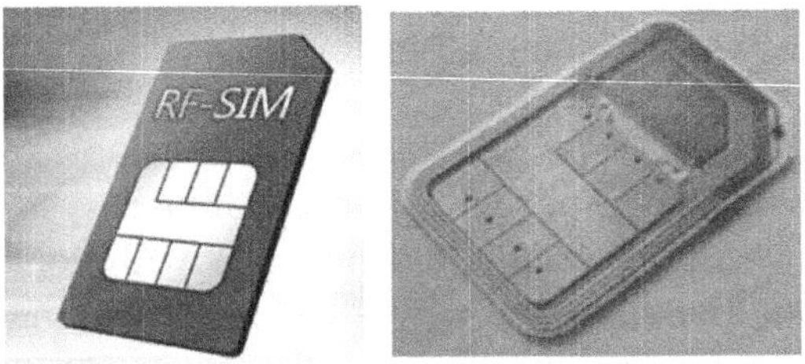

Figure 1. *Example of domains of use of ALM*

Reduction of H_min and low-power PCD

Over the past few years, it has sometimes been remarked upon that, in the field, many contactless chip cards (made of plastic) are more sensitive than the well-known threshold minimum field intensity H_min = 1.5 A/m. In addition, mobile phones and other NFC initiators (or, in terms of chip cards, PCD – proximity coupling devices), seeking to reduce their power consumption for reasons of battery life, have a great deal of difficulty in achieving H_min = 1.5 A/m. Consequently, there is a lot of pressure being applied (particularly to the ISO) to reduce the value of the transmitted field to approximately 0.8 A/m.

This is based on admirable sentiment but, in spite of these desires, in the name of interoperability with the immense existing range of chip cards/targets that conform fully to ISO 14443, it is absolutely impossible to implement such a reduction. One solution envisaged for the initiator would be for it to test whether it obtains responses at 1.5 A/m with targets present in the field and, after a few exchanges with those targets during which they communicate whether or not they are able to operate with a weaker field, begin to make energy savings.

Discussions are on the way at the ISO on this subject. The biggest issue is how to solve the problems of interoperability with existing devices.

NFC and RFID in HF

In radio frequency identification (RFID) – which, by definition, involves the traceability of items – in HF, at 13.56 MHz for many years, on the normative side, everything has been conscientiously constructed:

– a standard ISO 18000-3 mode 1, which draws heavily on the standard for vicinity contactless chip cards ISO 15693, is often applied for "Item management", when items are within the radii, on display cases, so with relatively short reading distances (50–70 cm). The principle of collision management employed for this standard performs far better than ISO 15693;

– a standard ISO 18000-3 mode 3 is an offshoot of the above standard, and its aim is to render an HF system at 13.56 MHz capable of supporting the same type of data encoding, EPCGlobal, as that used in ultra high frequency (UHF) by ISO 18000-6C;

– it must also be remembered that the NFC IP2 standard supports ISO 15693. Thus, there is not far to go before a mobile element NFC IP2 – such as a tablet, mobile phone, etc. – will be capable of reading EPCGlobal tags without too much difficulty, via ISO 18000-3 mod 3. In addition, the NFC Forum, through semi-conductor manufacturers (ST and NXP), is working on NFC Forum Tag 5 or "V", for vicinity, in compliance with ISO 15693.

Such is the situation as it currently stands.

The near future and beyond

However, on our planet, there are more than the proponents of contactless chip cards and mobile phones which work in contactless mode. There are also those who make electronic labels – and tags as well – in their billions each year, and not especially at 13.56 MHz.

NFC, RFID in HF and in UHF

Why, then, should we put two labels/tags on the same item – one in UHF to track it at medium/long distance (far field – a few meters) throughout the process of its manufacture and delivery, and another in HF (13.56 MHz) to track it from close by (near field) on the display unit? If we are able to identify it from afar using UHF and adjust to also use UHF to read it close by, we save on a tag. In this case, we merely need to give ordinary consumers "pocket" readers/interrogators using UHF to read that tag, and they can find out whether the product on display is a counterfeit, its origin, whether it was ethically manufactured, etc.

We are now back to a pocket item, such as a mobile phone, which needs to communicate this time using UHF over a short distance, in "near field" or "intermediary field". From here to speaking of "NFC or almost, in UHF", there is only one step which certain far-eastern friends have already taken, quite some time ago.

RFID, UHF and near-field Communication

Since 2007, a South Korean team at the ETRI – Electronics and Telecommunications Research Institute – has been focusing on the problem, and has reached a number of conclusions:

– On the market, in terms of items, sooner or later, there will be many UHF tags to serve the purpose of supply chain management.

– These tags will mainly be compatible with GS1 and EPCGlobal coding such as EPC C1 G2 – i.e. ISO 18000-6C compliant.

– No one would be pleased at having to put a second tag, at 13.56 MHz, on an article merely to duplicate the content so that the general consumers can read it with a "pocket reader" widely available, such as a mobile phone with NFC functions at 13.56 MHz in accordance with an IP1 or IP2 standard.

– If we wish to read the content of that UHF tag, then the mobile phone must be capable of reading something of the type ISO 18000-6x.

– As we only wish to read these tags at a short distance (8–15 cm) and do not want to affect the battery life of the mobile phones by overconsumption during data exchanges, it is preferable to use UHF rather than HF (a few mW in UHF transmission are sufficient to obtain a communication distance of some 10–30 cm), and where possible, we must use the phone's existing antenna.

– Thus, this leads us to use UHF RFID for an individual "mobile" device (be it a telephone or another kind of product), in near-field communication … so we can categorically state that these are NFC products. It is true; they do exactly what they say on the tin.

Of course, it is true that NFC at 13.56 MHz will continue on its happy way, with applications such as contactless chip card emulation where such cards are in widespread use (payment, transport, access control, etc.), but an "NFC UHF" version is by no means out of the question. Thus, for readers who want to escape tunnel vision as to "HF NFC", the following sections on "UHF NFC" should be helpful.

Standards pertaining to "mobile RFID"

A standard using conventional RFID frequency bands in UHF (860–960 MHz) has been drawn up, pertaining to "MIIM – Mobile Item Identification and Management": the ISO 29143 standard – *Information technology, Automatic identification and data capture techniques, Air interface specification for Mobile RFID interrogators* – which constitutes the ISO for the physical layer of "UHF NFC" systems for "mobile RFID".

This standard gives all the specifications for communication protocols for the "air interface" for general consumer mobile RFID interrogator systems (e.g. mobile phones, tablets, etc.) working with "passive/backscatter" tags or battery-assisted ones, and "Interrogator Talks First" (ITF) systems.

As is the case in normal RFID, the tags are remote-powered by the (electro)magnetic field generated by the interrogator's RF signal, and respond by modulating the reflection coefficient of their antennas – i.e. by "backscattering" the return data to the interrogator. The tags and tag-to-interrogator communication, therefore, are purely passive. Given the fact that multiple interrogators and multiple tags can communicate simultaneously, the standard describes how to manage carrier frequency collisions and data collisions between communications sent by the tags. In addition, the standard describes the way in which two interrogators can communicate with one another.

Convergences between HF and UHF NFC

In terms of applications, it is difficult to conceive:

– first, of end users one day having two different "mobiles": one for transport and banking applications and another to read product tags (to find their origin, usage instructions, warranty and so on);

– second, of having two completely distinct architectures for NFC functions in HF and UHF.

In the long run, this leads us to directly examine a possible convergence between NFC solutions, to work simultaneously in HF and UHF. Obviously, as mentioned above, it is advantageous to have the two NFC entities in one "mobile": first "HF NFC at 13.56 MHz" and second "UHF NFC at 900 MHz". Obviously, this somewhat complicates the manufacturing process when it uses discrete components for the two functions, because it is expensive and consumes a little too much power. On the other hand if the two are integrated, it is entirely feasible.

Common reception head for HF and UHF

Let us briefly examine the technical and technological problems posed by the integration of a common RF head for both HF and UHF NFC solutions, and the possible approaches to deal with these issues.

In terms of the reception and demodulation chain

The chain for reception and demodulation of the multi-band signal (HF to UHF – from 10 to 1,000 MHz) can be manufactured using an SDR architecture – software-defined radio – to decode and interpret the NFC IP1 and IP2 standards, ISO 18000 – 3 mod1 and mod3 at 13.56 MHz and ISO 18000 – 6 C compatible EPC C1 G2 and ISO 29143: mobile RFID in UHF. Additionally, the conventional technology CMOS is largely employed to create RF frontheads (amplification, mixers, FI, zero FI, etc.) – either HF or UHF.

The problem to be solved lies mainly in designing the architecture of an *HF/UHF mono-fronthead* for a reasonable price.

In the SDR part, all the processing is carried out in digital form, and mainly by software.

In terms of the antenna

To conclude this book, of course, we must return to the antenna.

The last thing we must do is to create an *HF/UHF mono-antenna* (gain, impedance matching, radiation diagram, etc.) and be able to work simultaneously in HF and UHF for near and far fields.

An example of a real-world mono-antenna is shown in Figure 2.

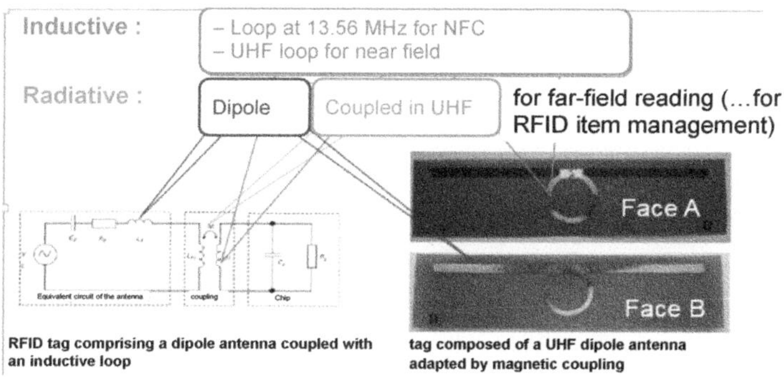

Figure 2. *Example of HF and UHF mono-antennas. For a color version of the figure, see www.iste.co.uk/paret/antenna.zip*

In fact, an HF/UHF monochip is linked to an astute combination of a dipole (for the radiative side for far fields in UHF) and a loop (for the inductive coupling side for near fields in HF and UHF). There are many other propositions that have been put forward by distinguished antenna designers all over the world. Readers should refer to their work (see Bibliography) for further details.

On all of this, we use a cryptographic mono-unit (small if possible, made of silicon and inexpensive – e.g. AES 128 or something similar) and return to Chapter 6 for applications of tags linked to the fight against counterfeiting, with T2 NFC tags using HF and UHF.

There are many fabulous technical achievements still to come, but in this field of activity, it is important to remember that we must always work from the principle that, if you do not do something yourself, someone else will do it instead, and thus come what may, the solution will eventually see the light of day.

We can always dream, so it is entirely possible that, in a few years' time, HF and UHF NFC devices using monochips, mono-antennas and monocryptography, which can also be used for applications with the *Internet of Things* (IoT), could well be a reality.

Perhaps, this could be the subject of my next book. Who knows?

Bibliography

Author bibliography

Glossary
ISO 19 762-3 – standardized vocabulary – radiofrequency identification (RFID).

Standards – Contactless chip-free cards
ISO 14 443 – 1 to 6 – 2015 edition – Proximity cards.
ISO 15 693 – 1 to 3 – 2015 edition –Vicinity cards.
JIS X6319 – 4 – High speed proximity cards – FeliCa.

Standards – NFC
ECMA 340 & 352.
ISO 18 092 – NCF IP 1.
ISO 21 481 – NFC IP2.

Standards – RFID
ISO 18 000 – 3 – 6C and – 63 – RFID, *Supply Chain Management.*

Standards – RFID Mobiles in UHF
ISO 29 143.

Standards – Contactless Communication for Public Transport
CEN 16 794 – 1 protocol.

Standards – Tests
ISO 10 373 – 6 – edition 2015 – conformity tests for proximity cards.
ISO 22 536 – conformity tests for NFC IP1 RF interface.
ISO 23 917 – conformity tests for NFC IP1 protocol.
CEN 16 794 – 2 conformity tests, public transport, contactless.

Standards – Privacy
Mandate M436 of the European Commission.
CEN EN 16 571 – *Privacy Impact Assessment* (PIA).

Standards – Marking of electrical and electronic equipment
EN 50419.

Proprietary standards
EMVCo L1 – contactless bank cards.
NFC Forum – Analog & Digital.

Regulations
ERC 70 03 – relating to the use of short range devices (SRDs).
ETSI 300 330 – conformity tests for radiofrequencies.
US code of federal regulations (CFR) Title 47, Chapter 1, Part 15, "Radio Frequency Devices".

"Human exposure" recommendations
ICNIRP International Commission on Non Ionizing Radiation Protection.

Publisher bibliography

[COK 12] COSKUN V., OK K., OZDENIZCI B., *Near Field Communication (NFC): From Theory to Practice,* Wiley, New York, 2012.

[PAR 03] PARET D., *Applications en identification radiofréquenceetcartes à puces sans contact*, Dunod, Paris, vol. 1, 2003.

[PAR 05] PARET D., *RFID and Contactless Smart Card Applications*, Wiley, New York, 2005.

[PAR 12] PARET D., *NFC (Near Field Communication). Principes et applications de la communication en champ proche*, Dunod, Paris, 2012.

[LAH 14] LAHEURTE J.M., RIPOLL C., PARET D. *et al., UHF RFID Technologies for Identification and Traceability*, ISTE, London and John Wiley & Sons, New York, 2014.

Index

A, B, C

D, E, F

H, I, J

L, M, N

P, R, S

T, U, V, Y

Other titles from

ISTE

in

Waves

2014

JARRY Pierre, BENEAT Jacques N.
RF and Microwave Electromagnetism

LAHEURTE Jean-Marc
UHF RFID Technologies for Identification and Traceability

SAVAUX Vincent, LOUËT Yves
MMSE-based Algorithm for Joint Signal Detection, Channel and Noise Variance Estimation for OFDM Systems

TING Michael
Molecular Imaging in Nano MRI

VALIÈRE Jean-Christophe
Acoustic Particle Velocity Measurements using Laser: Principles, Signal Processing and Applications

VANBÉSIEN Olivier, CENTENO Emmanuel
Dispersion Engineering for Integrated Nanophotonics

2013

BENMAMMAR Badr, AMRAOUI Asma
Radio Resource Allocation and Dynamic Spectrum Access

BOURLIER Christophe, PINEL Nicolas, KUBICKÉ Gildas
Method of Moments for 2D Scattering Problems: Basic Concepts and Applications

GOURE Jean-Pierre
Optics in Instruments: Applications in Biology and Medicine

LAZAROV Andon, KOSTADINOV Todor Pavlov
Bistatic SAR/GISAR/FISAR Theory Algorithms and Program Implementation

LHEURETTE Eric
Metamaterials and Wave Control

PINEL Nicolas, BOURLIER Christophe
Electromagnetic Wave Scattering from Random Rough Surfaces: Asymptotic models

SHINOHARA Naoki
Wireless Power Transfer via Radiowaves

TERRE Michel, PISCHELLA Mylène, VIVIER Emmanuelle
Wireless Telecommunication Systems

2012

LI Jun-chang, PICART Pascal
Digital Holography

2011

BECHERRAWY Tamer
Mechanical and Electromagnetic Vibrations and Waves
GOURE Jean-Pierre
Optics in Instruments

LE CHEVALIER François, LESSELIER Dominique, STARAJ Robert
Non-standard Antennas

2010

BEGAUD Xavier
Ultra Wide Band Antennas

MARAGE Jean-Paul, MORI Yvon
Sonar and Underwater Acoustics

2009

BOUDRIOUA Azzedine
Photonic Waveguides

BRUNEAU Michel, POTEL Catherine
Materials and Acoustics Handbook

DE FORNEL Frederique, FAVENNEC Pierre-Noël
Measurements using Optic and RF Waves

2008

FILIPPI Paul J.T.
Vibrations and Acoustic Radiation of Thin Structures

2006

BOUCHET Olivier *et al.*
Free-Space Optics

BRUNEAU Michel, SCELO Thomas
Fundamentals of Acoustics

GUILLAUME Philippe
Music and Acoustics

GUYADER Jean-Louis
Vibration in Continuous Media

Printed and bound by CPI Group (UK) Ltd, Croydon, CR0 4YY

13/09/2026

14961017-0005